# GIS AND PUBLIC DATA

## Bruce Ralston

**THOMSON**

**DELMAR LEARNING**

Australia   Canada   Mexico   Singapore   Spain   United Kingdom   United States

**THOMSON**

**DELMAR LEARNING**

# GIS and Public Data
## Bruce Ralston

**Vice President, Technology and Trades SBU:**
Alar Elken

**Editorial Director:**
Sandy Clark

**Senior Acquisitions Editor:**
James DeVoe

**Senior Development Editor:**
John Fisher

**Marketing Director:**
Dave Garza

**Channel Manager:**
Fair Huntoon

**Marketing Coordinator:**
Casey Bruno

**Production Director:**
Mary Ellen Black

**Production Manager:**
Andrew Crouth

**Production Editor:**
Stacy Masucci

**Technology Project Manager:**
Kevin Smith

**Technology Project Specialist:**
Linda Verde

**Editorial Assistant:**
Katherine Bevington

**Freelance Editorial:**
Carol Leyba, Daril Bentley

**Cover Design:**
Cammi Mosiman

This book is dedicated to Linda and Millie—
the two most fun women in my life.

# CONTENTS

# PREFACE

THIS TEXT AND COMPANION SOFTWARE are for GIS users who wish to take advantage of the wealth of public domain data available from federal agencies such as the United States Geological Survey, the United States Census Bureau, and the Environmental Protection Agency. Ideally, readers should have some familiarity with GIS concepts and terms, but the material should be accessible to students in introductory GIS courses.

Much of the data the United States government creates, maintains, and distributes is available at no or low cost, with many data sets available via the Internet. The chapters in the text–along with examples found in the companion PowerPoint slides–cover the content, scales, projections, and peculiarities users need to know in order to find, translate, and use these data sets.

The motivation for writing the text and companion software came from the author's own experiences in using and teaching others to use these data sets. The idea is to prevent tasks that should be straightforward and quick from turning into complicated, time-consuming chores, which is far too often the case in using data sets obtained from disparate sources. Knowing what you are doing and having access to the right software can save you days (or more) in lost time and avoid a great deal of frustration.

In the late 1990s the author began developing software for use with public domain data. The impetus for this book and its companion software was watching students struggle with what should have been relatively straight-forward tasks. The resulting software packages, all of which are included with this text, are used by individuals, private sector GIS users, GIS consulting companies, and local, state, and federal government agencies.

The text constitutes a single resource containing explanations of the major data sets available, reinforced by hands-on examples of each data set type in the form of widely available public domain GIS data sets. Each chapter starts with an overview of the data sets covered in that chapter, followed by in-depth exploration of the content, structure, and working details of each data set. Examples of commercial software (some of which was written by the author) in each chapter show you how to work with and get the most out of the data sets. Each chapter ends with a discussion of useful web resources available to the reader.

# ABOUT THE SOFTWARE AND THE COMPANION CD-ROM

This text's companion CD-ROM contains information referenced throughout the text. The content of the CD is organized into three folders: Software, PowerPoint, and Lookups. The Software folder contains software developed by the author. Translation routines for USGS digital line graphs and U.S. Census Bureau TIGER files are included, along with their user manuals. Extraction programs for working with U.S. Census Bureau Summary Files are also included. There are extraction routines for

Summary File 1 (state and advanced national files), Summary File 3 (state and national files), and 108$^{th}$ Congressional District Summary File 3 state files. Hyperlinked web pages of table and variable names are also included.

The PowerPoint folder contains files that provide information and examples referenced in the text. There is one file for each chapter. I have found that students find it useful to print out the PowerPoint slides so that they can take notes on them while they read or attend lecture. In addition, the PowerPoint slides allow for slide animation and the use of color graphics. By assigning some material to the PowerPoint slides, I have been able to limit the length, and hence the cost, of the text.

The Lookups folder contains database files referenced in the text. These files are useful in interpreting the meaning of coded variables used in several of the data sets covered. The tables were constructed by the author from information found in the technical documentation for the corresponding data sets.

# ABOUT THE INSTRUCTOR'S GUIDE

The Instructor's Guide CD contains supplemental material related to the topics in the book. Included on the CD are sample data sets, suggested exercises, and discussions of topics related to but not covered in the text. These topics include finding the name of the topographic quadrangle associated with a place; special datums and projections used in Puerto Rico, Alaska, and the Pacific; and other data sets of interest. Software updates and additional programs by the author are also included. These include software for extracting tables from Census 2000 summary files 2 and 4.

# ABOUT THE AUTHOR

Bruce A. Ralston is a professor and Head of Geography at the University of Tennessee. He has been involved in GIS since the mid 1980s, and has developed specialized software packages for the World Bank, the United States Agency for International Development, the World Food Programme, and the Southern African Development Council. He has also served as a GIS consultant for several private sector firms.Acknowledgments

There are several people to whom I owe thanks. Dr. Cheng Liu helped me develop my interest in GIS software. He also helped me write DLG2SHP and aided in the early versions of TGR2SHP. Dr. Shih-Lung Shaw has helped me stay up to date on the ever-changing world of GIS. Joseph Kreski of the United States Geological Survey made several important suggestions in his role as technical reviewer. His feedback is much appreciated.Catherine Miller of the United States Census Bureau's Geography Division patiently answered my questions on some of the new features and codes found in TIGER 2003. I also want to thank the team at OnWord Press. Daril Bentley, Carol Leyba, Jim DeVoe, and John Fisher have been productive, professional, and prompt. Thanks to them all for helping me complete this text.

# FEEDBACK

Readers with questions or feedback can contact Dr. Ralston via e-mail at *bralston@utk.edu*, or by mail at Department of Geography, University of Tennessee, 304 Burchfiel Geography Building, Knoxville, TN, 37996-0925.

CHAPTER 1

# PUBLIC DATA FOR GIS

## INTRODUCTION

SCIENTISTS, BUSINESS PEOPLE, RESEARCHERS, and others who work on the United States and make decisions related to location have access to a wealth of public domain data suitable for use in geographic information systems (GIS). Various agencies of the United States government build, maintain, and distribute such data at either no cost or at a nominal cost to the user. Estimates of government spending on geographic information vary widely. A conservative estimate by the National Academy of Public Administration put the value at $1 billion in 1997. However, this does not mean that such data are totally free or inexpensive.

Although many data sets are free in that they can be downloaded from the Web or found in public depository libraries, they can be expensive in the amount of time it takes to learn their formats, deal with projection issues, derive the information one needs, and get them into a GIS-usable format. In some cases, the data are distributed in formats that can be confusing or that require extensive translation into GIS-ready formats, or both.

There is technical documentation that explains the formats, and wading through the technical documentation is not something I discourage. However, you may not have the time or inclination to know all the details of, say, the Spatial Data Transfer Standard vector format or the content of Record Type A in the various versions of TIGER. Through many projects (and a lot of frustration), I have learned tricks and shortcuts, found useful software, and have written software to make using major public domain data sets easier. More than 70 percent of the typical GIS project is spent on gathering, formatting, and manipulating data (as studied by Aronoff, 1989, and others). The goal of this text is to pass on this information to you, the reader, so that data-related tasks take minutes instead of hours or days.

You may wonder why a text like this is needed. After all, there is a lot of documentation available on the base data sets produced by government agencies. Much of it is very detailed and often technical. However, when reading the government documentation, I often feel like the unfortunate fellow who asks the watchmaker for the time. The watchmaker then goes into a long discussion of just how a watch is made! For the uninitiated, the expense of using these data often outweighs their benefit. In addition, users often try to use these data sets in inappropriate ways (e.g., predicting effects at 1-meter resolution using data with 30-meter resolution). Put simply, there is no free map! (See slides 1 through 5.)

Slides 1–5

# TOPICS COVERED

This opening chapter covers some mapping basics, base data sources, and some common places where one can obtain the data described in this text. The remaining chapters explore the base data sets in detail (see slides 6 through 8). Chapter 2 focuses on digital raster graphics

(DRGs) and digital orthophoto quads (DOQs). DRGs are digital images of topographic maps (see slides 9 through 12). You are probably familiar with topographic maps. These maps are available at different scales, with different projections and datums. A common extent for such maps is the quadrangle (meaning "four angles," often called a quad).

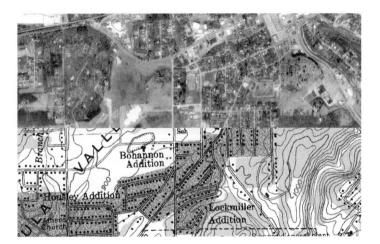

Slides 6–12

At its most detailed scale, 1:24,000, a quadrangle covers 7.5 minutes of latitude and 7.5 minutes of longitude. About 55,000 of these "seven-and-a-half-minute" quadrangles at this scale are needed to cover the United States.

Topographic maps are updated primarily using information derived through the National Aerial Photography Program (NAPP). The NAPP is discussed in more detail later in this chapter. Suffice it to say that aerial photos produced by the NAPP (among others) are used to update topographic maps. Aerial photos are processed to produce digitally rectified images of 3.5-minute by 3.5-minute sections of a topographic map. For a 7.5-minute quad, this represents one quarter of the quad (thus the name digital ortho quarter quad, abbreviated DOQQ). Figure 1-1 shows a DOQQ atop a DRG for the 1:24000 Athens, Tennessee, topographic map.

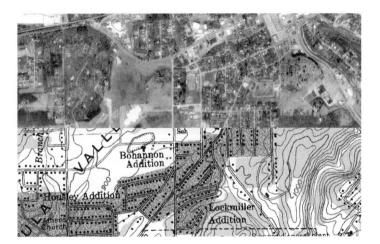

Fig. 1-1. A DOQQ and corresponding DRG.

Chapter 3 focuses on vector representations (digital line graphs, or DLGs) of topographic maps. DLGs take the information on a topographic map and break it into nine different vector layers. As you will see in later chapters, DLGs serve as the source for several other digital data sets to be discussed (see slide 13).

Slide 13

Chapter 4 covers digital elevation models (DEMs), the recently created National Elevation Data (NED) set, and Shuttle Radar Topography Mission (SRTM) data. All of these data sources contain information about elevation at locations (see slides 14 and 15). The DEM and NED contain the same information, but they use different coordinate systems, projections, and datums. SRTM data, as the name implies, is based on information gathered by a Space Shuttle mission. This is the only data set considered in this text that contains coverage outside the United States and its territories. Figure 1-2 shows a DRG draped over a DEM for Mount Rainier.

Slides 14–15

Fig. 1-2. Draping a DRG over a DEM.

Chapter 5 covers vector and raster representations of land use and land cover (see slides 16 through 18). The Land Use Land Cover (LULC) data is in vector format, whereas the National Land Cover Data (NLCD) consist of

Slides 16–18

raster information. Not only are LULC and NLCD data in different formats, they differ in timeliness (NLCD is more up to date), sources, coordinate systems, datums, and projections. They also use slightly different land cover classification systems.

Slide 19

Chapter 6 covers the widely used Topologically Integrated Geographic Encoding and Referencing (TIGER) files (see slide 19). These files serve as the basis for most geocoding in the United States, and they are required if you are to make thematic maps of U.S. Census data, such as maps of home ownership, the median age of a population, ethnicity, or other population or housing characteristics. Although TIGER may not be as spatially accurate as some other data sets, its importance in understanding data produced by the U.S. Census Bureau, such as the Census of Population and Housing, cannot be overestimated. Unlike many of the other data sets you will read about, these files have a unique data format, and that format changes with each new release of TIGER.

The data collected for the 2000 Census of Population and Housing is distributed in summary files, which are the focus of Chapter 7. These summary files contain a wealth of data, but they can be quite large. They may even overwhelm some commonly used database software. The information derived from the summary files, once processed, can be directly tied to the geographic census units in TIGER (see slides 20 through 22), allowing for the mapping of populations and housing characteristics.

Slides 20–22

Chapter 8 deals with the most complex form of vector data covered. Whereas other vector data sets used simple topology—points, lines, and polygons—the National Hydrography Dataset (NHD) contains composite elements. In particular, route systems and region systems (i.e., which watershed a stream, for example, belongs in and what direction it is flowing) play a major role in the

NHD. In addition, there are numerous files that reflect the relationships among elements of the surface drainage network. The NHD differs from other data sets considered in that the United States Geological Survey (USGS) has developed sophisticated tools for handling the NHD in ArcView 3 and ArcGIS 8 (see slide 23).

Slide 23

In addition to the topics outlined previously, you will learn about using several software packages with these data sets. These packages are:

- ❏  Global Mapper (Global Mapper Software, LLC)

- ❏  ArcCatalog, ArcMap, ArcToolbox, Workstation Arc-Info, and ArcView 3 (Environmental Systems Research Institute, or ESRI)

- ❏  TGR2SHP, SF1toTable, and SF3toTable (author)

- ❏  DLG2SHP (written by Dr. Cheng Liu and author)

- ❏  NHD ArcView Toolkit (USGS)

You will also learn about several web services that provide data, often free of charge. You will see how to use the Census Bureau's web pages to extract information from summary files. This is a reasonable strategy if you need just a few variables for a single summary level and limited area, such as census tracts in a particular county.

> **NOTE:** *The information in this book is current at the time of its publication. However, the Internet is a dynamic medium, and web addresses and interfaces may change. In addition, the policies of groups such as the USGS, EPA, and the Census Bureau can change.*

Each of chapters 2 through 8 contains a table summarizing the major points associated with the data sets discussed in the chapter. Information such as map scales, data formats, projections, software considerations, software discussed in the chapter, and other items of interest are listed. These tables are meant to provide you with an "at-a-glance" summary of information on each data set.

# DATA SOURCES OFTEN OVERLOOKED

The sources of data described in the text consist primarily of web sites run by government agencies, such as the Census Bureau or the USGS. However, there are other sources you may wish to consider. Some sources provide the files free of charge, whereas others charge some fee, with the rate dependent on the distribution media used. Usually, it is cheaper to download data than to receive a CD.

## Map Libraries

Many map libraries, particularly those that are part of a Federal Depository Library, have digital spatial data available on CD. You can copy the files you need from those CDs, as most federal data sets are in the public domain. There are many of these. As an example, an Internet search on *Map Library DRG* resulted in over 10,000 entries (see slides 24 and 25). Copying files from the CDs is usually straightforward.

Slides 24–25

## State User Group Web Sites

Most states have web sites where users can download DRGs and other base spatial data. The California Spatial Information Library is one such source. Another is the Tennessee Spatial Data Server. A web search for your state may yield a source of free public data sets. Often these data sets have been translated into a GIS-ready format and have been reprojected, usually into State Plane coordinates. The URLs for some selected sites are in the Notes section of slide 26 of the PowerPoint slides.

Slide 26

## Commercial Data Sources

Several companies have set up web sites where users can order digital data. The USGS has contracted with some private companies, its business partners, to distribute some of the data sets the USGS creates, such as the Spatial Data Transfer Standard (SDTS) raster format DEMs (see slide 27).

Slide 27

# MAPPING BASICS

Before dealing with the data sets described previously, there are a few basic concepts and data sources that should be covered. The discussion here is designed to provide a brief overview of concepts discussed in the remaining chapters.

## Projections and Datums

Earth is a 3D object, a spheroid. Rendering any 3D object, in this case the earth, in two dimensions requires using some form of mathematical transformation. Such transformations are called map projections. A map projection takes latitude and longitude measurements and transforms them into a 2D coordinate system. The USGS web site *(http://erg.usgs.gov/isb/pubs/Map Projections/ projections.html)* explains some of the details of the most commonly used map projections (see slide 28). Some of the more common projections you will encounter are Universal Transverse Mercator (UTM), Albers Equal Area projection, and State Plane projections. One widely used form of reporting is called geographic units. This is not really a projection. It is merely a form of reporting latitude and longitude as if they were 2D coordinates.

Slide 28

UTM projections are in a class called cylindrical projections. Each UTM projection has a zone number associated with it, with each zone spanning 6 degrees of longitude. That is, UTM is appropriate for areas that are 6 degrees (or less) of longitude in width and that lie between 84 degrees north latitude and 80 degrees south latitude. UTM is a common projection for topographic maps and data sets based on topographic maps. Angles and shapes within a zone have minimal distortion because at any point the scale is the same in any direction. Such projections are called conformal. Distances at the edges of a zone are subject to a scale factor of 0.9996. This is quite accurate.

Albers Equal Area projection is a conic projection that is often used to map the conterminous United States. As the name implies, relative areas are preserved under this type of projection. Albers is used to construct national-level data sets such as the NED and the NLCD.

Each state has at least one State Plane coordinate system. Large states, particularly Alaska, have several State Plane coordinate systems. Projection parameters will vary by state, but there are only three types of projections used: Lambert Conformal, Transverse Mercator, and Oblique Mercator. The Lambert Conformal projection is often used for states that have an east-west orientation, the Transverse Mercator for those with a north-south orientation, and the oblique for states that have an oblique orientation.

No matter what projection (including geographic units) is used, an assumption has to be made about the shape and size of the earth. It is usually assumed that the earth is close to a sphere, a spheroid. That is, it is assumed that an approximation of the earth's shape can be obtained by spinning an ellipse about its minor axis. The resulting shape is a spheroid with semi-major (equatorial)

Slides 29–30

and semi-minor (polar) axes. This is an approximation, as the earth is not a perfect sphere. There have been many attempts to measure the lengths of the axes. Each measurement is part of what is known as a datum (see slides 29 and 30).

Because the earth is not a perfect spheroid, it makes sense that different estimates of the semi-major and semi-minor axes will work better in some parts of the earth than in others. Indeed, there is a long history of geodetic measurement.

> **NOTE:** *For more details on geodetic measurement, see* Geodesy for the Layman *at* http://www.nima.mil/GandG/geolay/toc.htm.

A datum is needed to estimate latitude and longitude. Even when map coordinates are not projected, a datum still influences the estimates of latitude and longitude. This is why you can take two maps of the same area in the same units at the same scale and they may not register (that is, they can be based on different datums).

As you will see in Chapter 4, the National Elevation Dataset and the Shuttle Radar Topography Mission data both supply information for elevations, but they are based on different datums. As a result, when comparing the two data sources you can discern a slight shift in locations. For the uninitiated, this difficulty in getting maps to "register" can be quite frustrating. The major datums used in the data sets you will study include:

❏ North American Datum, 1927

❏ North American Datum, 1983

❏ World Geographic System, 1984

In addition, areas such as Puerto Rico, Alaska, and Hawaii often use locally correct datums.

## Scale and Resolution

Two related but distinct concepts are cartographic scale and resolution. The scale of a map refers to the ratio of map units to ground units. For example, National Aerial Photography Program (NAPP) photos are produced at a 1:40000 scale. That is, one unit of distance on the photograph represents 40,000 units on the ground. The photos at this scale are 9 inches by 9 inches. Put another way, the photograph represents an area of approximately 360,000 square inches, or 30,000 square feet, or 5.68 square miles. The USGS provides these photographs at other scales. For example, you can purchase an NAPP photo that is 36 inches square. In this case, the scale is 1:10000 (see slide 31).

Slide 31

Photographs consist of pixels, the individual "picture elements" of gray shades or tones of color. The resolution of NAPP photos is 1 meter. That is, 1 meter is the smallest area on the earth's surface a pixel represents. We cannot discern features that are less than 1 meter across on the image, because the pixel size is 1 square meter. Whether at scale 1:40,000 or 1:10,000, the resolution of the photograph is still 1 meter. Put another way, for photographic data we can change scales but not resolution.

## Precision and Positional Accuracy

It is possible to be quite precise but very inaccurate (see slides 32 and 33). That is, precision and accuracy are not synonymous in the arena of geographic study. Precision refers to the number of significant digits for which location measurements are taken. TIGER, for example, reports coordinates in decimal degrees to six decimal places. This may seem quite precise. However, the original positional accuracy of TIGER is only as good (and may be worse) than that found in 1:100000

Slides 32–33

USGS topographic maps. According to the Census Bureau, the positional accuracy of TIGER data is "not suitable for high-precision measurement."

Positional accuracy refers to how well map coordinates match known coordinates of objects. The degree of positional accuracy in a data set usually reflects the accuracy of the source information, errors created in data transformation and projection, and (perhaps most importantly) the accuracy required for the intended use of the data set. For 1:24000 DLGs, positional accuracy may be quite important.

If a data user is concerned about positional accuracy, 1:24000 DLGs are the preferred data format. If a data user were more concerned about having current data, about being able to geocode addresses, or about creating neighborhood population maps, TIGER files would be the preferred data format.

# IMPORTANT DATA SOURCES

The various data sets discussed in the following chapters all have a lineage. That is, they are derived from different but related input-data sources. Two sources commonly used as input to public domain spatial data are the NAPP photographs and the Geographic Names Information System (GNIS). It is important that you are aware of these information sources.

The NAPP program involves taking aerial photographs of the conterminous United States and Hawaii at five- to seven-year intervals. The resulting photographs are used to update topographic maps and, by extension, all data sets derived from them, such as DLGs and DEMs. NAPP photos, once digitally rectified, are used to create DOQQs, which are discussed in Chapter 2.

NAPP photos are taken at specified points along a specific flight path. 7.5-minute topographic maps are divided into quarters. Flight paths are then plotted from north to south down the center of each eastern and western half of the quad. Photos are taken at the northern and southern edges of each quarter and in the center of each quarter. Figure 1-3 illustrates flight lines and points of photography (see slides 34 through 36).

Slides 34–36

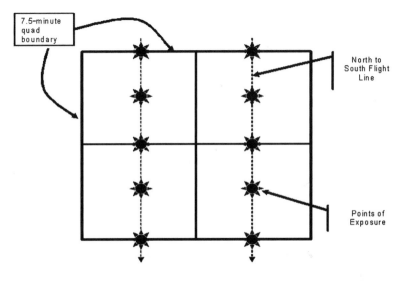

Fig. 1-3. NAPP flight lines and points of photography.

Two images are taken at each point of exposure: one with the tree's leaves on and one with leaves off. The images are taken at an altitude of 20,000 feet on cloudless days. The resulting images are not maps. Distortion resulting from topography and camera angle is removed by draping the aerial images over a DEM and making adjustments so that positional accuracy is enhanced. The result is a photograph that can be read and used like a map.

Many data sets contain names of features, such as place names, airport names, and so on. These names are usually taken from the GNIS, which is the federal database of

Slides 37–42

official place names. It contains nearly two million physical and cultural features. Name fields in data sets such as DLGs, NHD, and populated places and county divisions in TIGER will be filled with names that come from the GNIS. The GNIS web site (*http://geonames.usgs.gov*) supports user queries (see slides 37 through 42).

# WEB SITES AND DOCUMENTATION

Federal public domain databases are usually well documented and this documentation is often available over the Internet. Useful web sites exist for GNIS data, NAPP data, and National High Altitude Photography (NHAP) data (see slide 43). The NHAP is a precursor of the NAPP. The purpose of this section is to make you aware of this documentation and to provide a sense of the types of information available. Documents describing various USGS products can be found by clicking on the Products list at *http://edc.usgs.gov* (see figure 1-4).

Slide 43

Figure 1-5 shows the possible links to the various aerial products. Descriptions exist of the NAPP and NHAP, along with other aerial products, such as satellite photography, aircraft scanning, and special series such as Antarctica and DOQs (see Chapter 2).

The NAPP product page, *http://edc.usgs.gov/products/aerial/napp.html*, contains a product description, pricing information, instructions on searching and ordering NAPP products, instructions on obtaining custom enlargements, and certification information (see figure 1-6). Prints of NAPP photos are available at various scales, and in black-and-white and color. The details are listed in the Prices section of the product page. More technical information on the NAPP is available via links in the Product Description section.

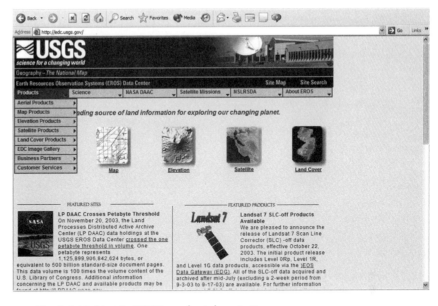

Fig. 1-4. Linking to USGS product descriptions.

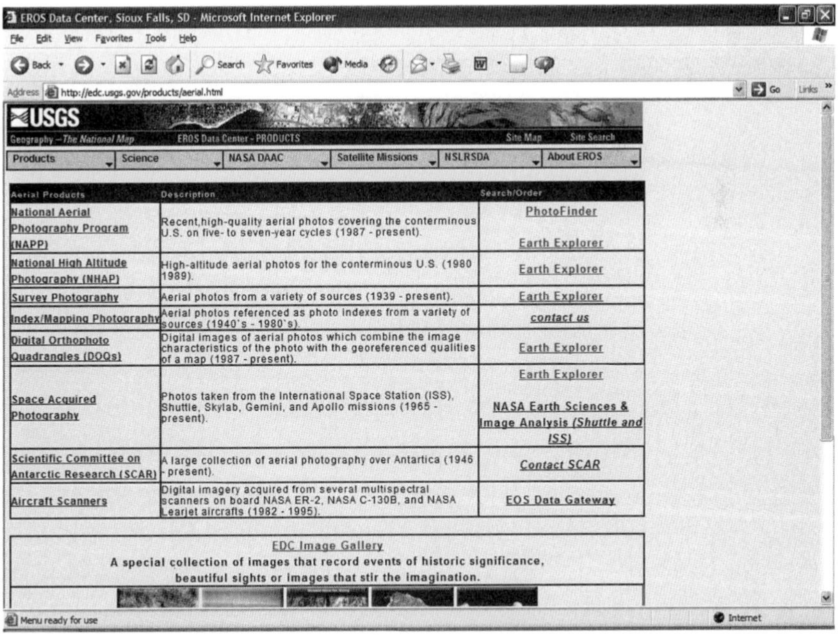

Fig. 1-5. The aerial products available from the USGS.

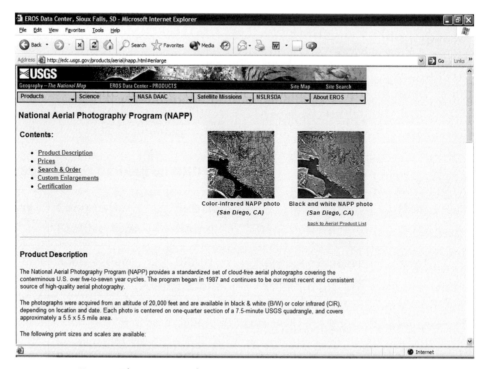

Fig. 1-6. The NAPP product page.

The technical information available for the NAPP includes the following.

❏ NAPP Status Maps (*http://mcmcweb.er.usgs.gov/ status/napp_stat.html*). These maps indicate the schedule and location of aerial photography. It is possible to retrieve PDF files of the NAPP status by state that indicate the availability of black-and-white and color infrared photography.

❏ Camera calibration information (*http://erg. usgs.gov/tsb/osl/smaccompen.pdf*). This document describes USGS aerial camera specifications, and includes instructions and fee information for having cameras calibrated so that they meet NAPP standards. Technical procedures for compensating for camera lens distortion are described, including the mathematical transformation used.

❑ NAPP Guide (*http://edc.usgs.gov/guides/napp. html*). This guide includes information on the background, coverage extent, spatial resolution, data organization, and procedures for obtaining NAPP products. It is more detailed than the product page (shown in figure 1-6). Technical information includes film specifications (both black-and-white and color infrared) and flight specifications.

The NHAP has pages similar to those for the NAPP. The NHAP, which was phased out in 1989 and replaced by the NAPP, was developed to support land-use/land-cover mapping (see Chapter 5). The product page for the NHAP contains the same five sections found at the top of the NAPP product page (see figure 1-6). As with the NAPP, more detailed information is available in the product guide (*http://edc.usgs.gov/guides/nhap.html*), shown in figure 1-7.

Slide 44

A general description of selected USGS products can be found at the Digital Backyard web site (*http://mapping. usgs.gov/digitalbackyard*). This site also contains a description of the NAPP, but it is geared to a less technical audience than that found at the NAPP product page (see slide 44). The Digital Backyard is part of the TerraServer site (described in Chapter 2). A list of sites containing online aerial photo images and maps is available at *http: //mapping.usgs.gov/partners/viewonline.html* (see slide 45).

Slide 45

GNIS information is available from the GNIS home page at *http://geonames.usgs.gov/gnishome.html*. The links section of that page contains a link to the GNIS Data User's Guide (see figure 1-8), which contains information on the

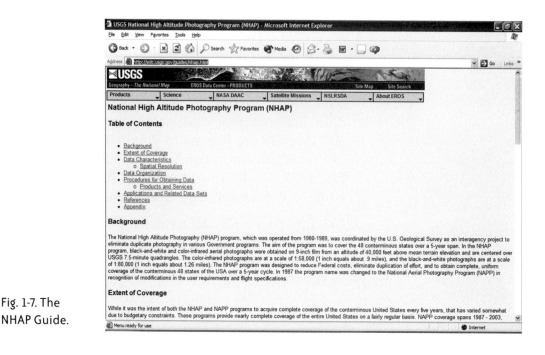

Fig. 1-7. The
NHAP Guide.

three databases that make up the GNIS system: the
National Geographic Names Data Base, the USGS Topographic Maps Names Data Base, and the Reference Data
Base. Descriptions of the record layout of each of these
databases are included. A gazetteer download page (*http:
//geonames.usgs.gov/stategaz/index.html*) allows users
to download information about states and territories.
There are four special extracts from the GNIS system that
can be downloaded. These are:

❒ *U.S. Populated Places* file, which contains information on communities throughout the United States

❒ *U.S. Concise* file, which contains information on
major physical and cultural features

❒ *Historical Features* file, which lists information on
places no longer in existence

❒ *Antarctica* file, which lists features and names of
Antarctica

One of the links from the *GNIS Data User's Guide* leads to a description of FIPS55 data (*http://www.itl.nist.gov/fipspubs/fip55-3.htm*). FIPS stands for Federal Information Processing Standard. FIPS values are crucial to a full understanding of databases developed by the U.S. Census Bureau. You will read more about FIPS codes and how they are used in GIS in chapters 6 and 7.

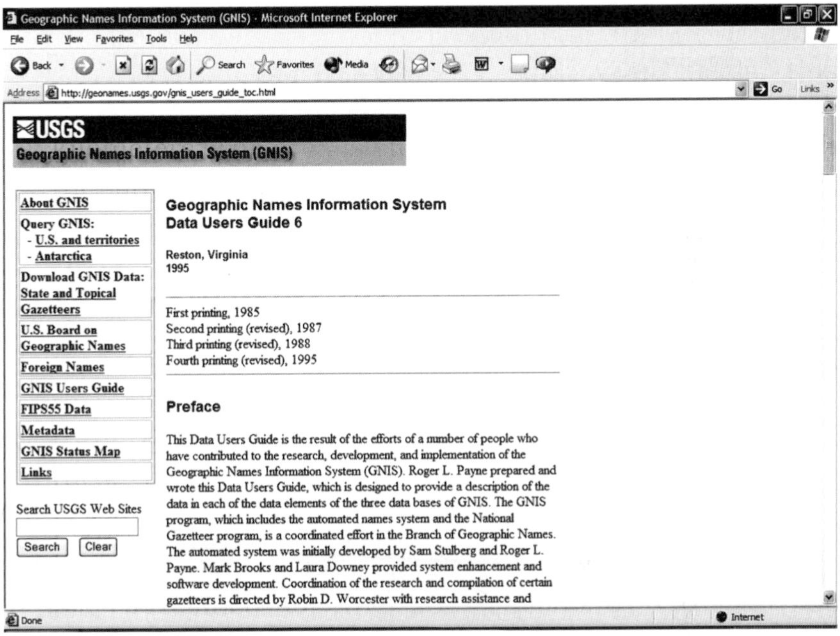

Fig. 1-8. The *GNIS Data User's Guide*.

# CHAPTER 2

# DIGITAL RASTER GRAPHICS AND DIGITAL ORTHOPHOTOS

## INTRODUCTION

DIGITAL RASTER GRAPHICS (DRGs) are scanned images of topographic maps. Digital orthophotos (also known as digital orthophoto quads or digital ortho quarter quads— DOQs or DOQQs, respectively) are corrected digital versions of aerial photographs that can be used like a map. Both are useful as background layers in GIS, for placing information in context, and for locating data collected in the field. Most graphics and image viewing software will display these files, but to fully take advantage of the geographic referencing information that comes with them you will need software that accommodates map projection information. These files may be in different map projections, and therefore care must be taken when combining files that have different projections and when reprojecting files to a new coordinate system. Table 2-1 presents an overview of these data sets.

**Table 2-1: DRG and DOQ Overview**

| Data Set | DRGs | DOQs |
|---|---|---|
| Description | Scanned images of topographic maps. | Aerial photographs that have been rectified and saved in a map projection, usually UTM. |
| Format | Usually distributed as GeoTIFF files with world and metadata files. | Distributed as GeoTIFFs with world and metadata files. Also available in "native" format and in MrSID format. If ordered on CD-ROM, files are compressed JPGs. |
| Scales | Depends on source maps. Usually match scale of corresponding quad. Common scales are 1:24000, 1:100000, and 1:250000. | 1:12000 for quarter quads and 1:24000 for 7.5-minute quads. |
| Projection/ Coordinate System | Varies by area. Usually UTM in meters based on NAD27 or NAD83. | Varies by area. Usually UTM in meters based on NAD27 or NAD83. |
| Original Source | USGS topographic maps; some orthographic and planimetric maps. | National Aerial Photography Program (APP) images. |
| Software Considerations | Can be viewed with any TIFF-enabled software. Need GIS-based software to properly geo-register and project. | Can be viewed with any TIFF-enabled software. Need GIS-based software to properly georegister and project. |
| Other Considerations | Care must be taken when projecting or combining across UTM zones. Edges of adjacent areas may not match. | Care must be taken when projecting or combining across zones. |
| Software Used in This Chapter | Global Mapper, ArcCatalog, ArcMap. | Global Mapper, ArcCatalog, ArcMap. |

# DRGs

You have probably worked with topographic maps, often referred to as "topos." The USGS (United States Geological Survey) and other organizations, such as the Tennessee Valley Authority (TVA), produce these paper maps. Producing topographic maps of the conterminous United States, Puerto Rico, and Hawaii at 1:24,000 scale and Alaska at 1:1,000,000 scale required enormous effort. The production of topographic maps started in 1879, and a full set of 55,000 such maps was not completed until 1990. The entire effort took more than $1 billion and 33 million man-hours, of labor (Kelmelis, John A., et al., "The National Map: From Geography to Mapping and Back Again," *Photogrammetric Engineering and Remote Sensing*, vol. 69, no. 10, Oct. 2003, pp. 1109–1118).

Starting in 1995, the USGS, TVA, and the California Spatial Information Library began the process of scanning and producing digital images of these maps (see slides 1 through 3). The resulting products are called DRGs, which are *digital* because they are scanned versions of maps, *raster* because they are in raster format (a grid of pixels), and *graphics* because they come from cartographic products (i.e., paper topographic maps). An example of a DRG is shown in figure 2-1.

Slides 1–3

A DRG is a digital image of a paper map. The original DRGs were produced by scanning maps at 250 dots per inch (dpi). As scanner technology has improved, and disk drives are more capable of handling increased file sizes, the density of the scan has increased. In May of 2001, the USGS announced that it would support scans of up to 1,000 dpi. This can yield high-quality maps that do not become over-pixilated as one zooms in on the map image. It also requires more disk space to store the high-resolution maps. DRGs created before October of 2001 are at

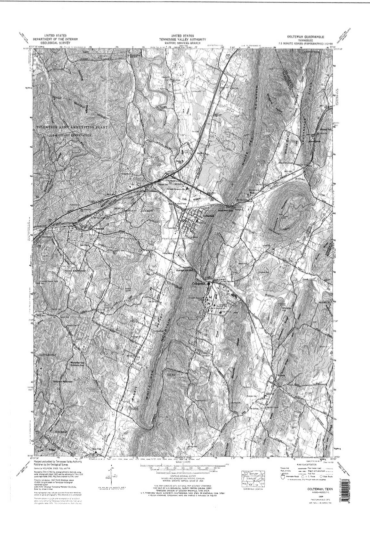

Fig. 2-1. A 1:24000, 7.5-minute DRG for the Ooltewah quadrangle.

250 dpi. Since that time, most DRGs are at 500 dpi, but because production of new DRGs has been slow, the bulk of USGS DRGs is still 250 dpi. Many state sites and private companies, however, have created their own higher-resolution DRGs.

Because DRGs are scanned versions of topographic maps, it is useful to have some knowledge about those maps. Topos are produced in map series, and each series is usually named after the number of minutes (as in degrees-minutes-seconds, of latitude and longitude, not time) the series covers. Common series include 7.5-minute quadrangles (usually written 7.5′), 7.5-minute by 15-minute (7.5′ x 15′) quadrangles, 30-minute by 60-minute (30′ x 60′) quadrangles, 1-degree x 2-degree (1° x 2°) quadrangles, and other formats for Alaska. For a listing of topographic map series and scales (including those for Alaska, Puerto Rico, and Antarctica), see *http://erg. usgs.gov/isb/pubs/factsheets/fs01502.pdf.* Slide 4 shows the relative sizes of the 7.5′ x 7.5′ DRG, the 30′ x 60′ DRG and the 1° x 2° DRG. Most DRGs are based on topographic maps, but there are also orthographic and planimetric DRGs. Orthographic DRGs are a combination topographic map and composite aerial photograph. Planimetric DRGs do not contain contour lines.

Slide 4

For most of the United States, DRGs are in UTM coordinates, one of the coordinate systems shown on topographic maps. However, California DRGs are available in Albers Equal Area projection. The Tennessee Valley Authority (TVA) distributes its DRGs in either the original UTM-meters format or in Tennessee State Plane coordinates, 1983 datum in feet. (The Tennessee State Plane projection is a Lambert Conformal Conic projection.) Other special series projections and scales exist for Puerto Rico, Alaska, Hawaii, and islands in the Pacific. The coverage of DRGs and the organizations responsible for creating them are indicated in figure 2-2.

Each DRG usually consists of three files: the data file, a metadata file, and a World file. The data file is usually a TIFF format and has the extension *.tif.* The metadata file contains information about the data, such as who produced it, its scale, the date, the accuracy standard, and so on. Metadata files usually have an extension of *.fgd,* and the World file has an extension of *.tfw.* The World file is used to describe the size of each pixel (in map units), the rotation angle, and the coordinates of the upper left-

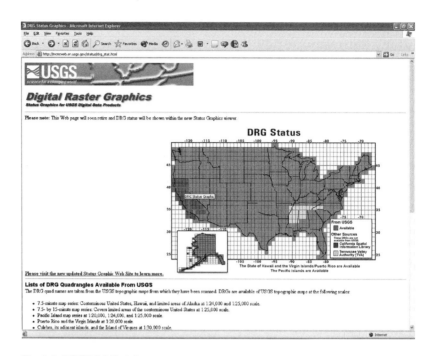

Fig. 2-2. USGS DRG status map.

hand corner of the map. Without the World file, the GIS software does not know what the spatial coordinates are of the DRG, and the DRG will therefore not overlay with other data registered to that area.

Exceptions occur with the two other major creators of DRGs. TVA gives its metadata file an extension of *.met*. California does not include a metadata file, but does include projection information in the TIFF header. Prior to December 2003, California DRGs put projection information in an ESRI projection file (extension *.prj*), which contains map projection details. Both TVA and California DRGs have TFW World files. As is the case with other image files, DRG files need to have the same base name to be properly georegistered in a GIS. For example, a DRG, World file, and header file for Sheboygan all need to have the base file name *Sheboygan*.

Slide 5

TIFF (Tagged Image File Format) files are one of the most widely used formats for distributing images (see slide 5). However, they were not developed specifically to handle spatial data, but for images such as pictures and other graphics. A TIFF contains space for tags (hence the name) in which to store information about the image. A special form of TIFF file is the GeoTIFF, which contains projection information in the header.

> **NOTE:** *The details of the GeoTIFF format can be found at* http://www.remotesensing.org/geotiff/spec/geotiff2.4. html.

The tag information will allow "TIFF-literate" software to know the size of each pixel in map units and the extent of the image. Thus, most GIS software should be able to display the image and georeference it correctly.

The TIFF file by itself does not contain all the information needed to work directly with the map. For example, information about the map lineage might not be stored in the header. (Lineage refers to the processing data goes through as it is being produced.) Thus, DRGs (and other spatial image files) often have a corresponding metadata file that contains information about the map. This would include the lineage, source data, projection type, datum, any false eastings and northings, and the units of measurement (such as feet or meters).

The World files contain information on the dimensions associated with each pixel, the rotation angles of the file, and the coordinates of the center of the upper left-hand pixel. For DRGs, the World file corresponds to the mapped area, not the entire paper sheet.

Consider, for example, the 30-minute x 60-minute DRG for Sleep Hole Mountains, California. This DRG is named

F34115A1. The World file *(F34115A1.tfw)* contains the following information.

*10.1600000000*

*0.0000000000*

*0.0000000000*

*−0.1600000000*

*367106.9214457296*

*−78769.1252430404*

This indicates that each pixel's width represents 10.16 map units (in this case, meters) and its height represents −10.16 units. The last two numbers represent the center of the upper left-hand pixel. The TIFF header will indicate the number of rows and columns in the image. Thus, the drawing software can determine the extent of the map. For this DRG, there are 5,884 rows and 9,313 columns. The corresponding map has the following coordinates for the center of the bottom right-hand pixel:

```
X = 367106.9214457296+10.16*(9313-1) = 461716.01446
Y = -378769.1252430404-10.16*(5884-1) = -438540.4052
```

You might wonder why the metadata is necessary. After all, the TIFF file contains information that allows software to georeference the image. If you simply want to display the image, then the metadata file is not needed. However, if you want to project the image into another coordinate system, knowledge of the existing projection system is crucial. For most DRGs, that information is kept in the metadata file. If you want to know the lineage of the image, you must use the metadata for that also.

Slides 6–9

You might have noticed that the file referenced previously had a strange name, *F34115A1*. There is a method to the naming convention of DRGs (see slides 6 through 9). The name of each DRG can be broken into four parts. The first part is a letter, in this case *F*. The letter indicates the map series on which the DRG is based. This will indicate the map scale and class. The meaning of the letters that start DRG file names are provided in Table 2-2 (source: *http://topomaps.usgs.gov/drg/drg_name.html*).

**Table 2-2: Meaning of Leading Letter**

| Category | Series | Scale | Class |
|----------|--------|-------|-------|
| R | 7.5-minute | 1:20,000 | Topographic |
| O | 7.5-minute | 1:24,000 | Topographic |
| P | 7.5-minute | 1:24,000 | Orthographic |
| L | 7.5-minute | 1:25,000 | Topographic |
| J | 7.5-minute | 1:30,000 | Topographic |
| K | 7.5-minute x 15-minute | 1:25,000 | Topographic |
| I | Alaska | 1:63600 | Topographic |
| G | 30-minute x 60-minute | 1:100,000 | Planimetric |
| F | 30-minute x 60-minute | 1:100,000 | Topographic |
| C | 1-degree x 2-degree | 1:250,000 | Topographic |

The file in this example starts with the letter *F*, and thus it is derived from a 1:100,000 topographic map. The next five digits indicate the latitude and longitude of the southeast corner of the map. The first two digits are the latitude, and the last three are the longitude. For the sample file from California, you can see that the southeast corner has the coordinates 34 degrees north latitude and 115 degrees west longitude.

The final part of the file name refers to the map index number. How these are determined depends on the map series.

> **NOTE:** *DRG file-naming conventions are described at the Topomaps web page found at* http://topomap. usgs.gov/drg/drg_name.html.

As regards 7.5-minute quads, consider a block of latitude and longitude 1 degree by 1 degree. For the purposes of illustration, this is displayed as a square in figure 2-3. This 1-degree by 1-degree cell is broken into an 8 x 8 grid. Because there are 60 minutes in one degree, each cell in the resulting grid is 7.5 minutes on a side. The columns are labeled from 1 to 8 from east to west. The rows of the resulting grid are labeled A to H from south to north. Consider a DRG named *O36084a1.tif.* From this name, you can see that this DRG is for a 7.5-minute topographic map with southeast corner at 36° North and 84° West. The western-most extent of the DRG is 84 degrees and 7.5 minutes west, and the northern-most extent of the DRG is 36 degrees and 7.5 minutes north.

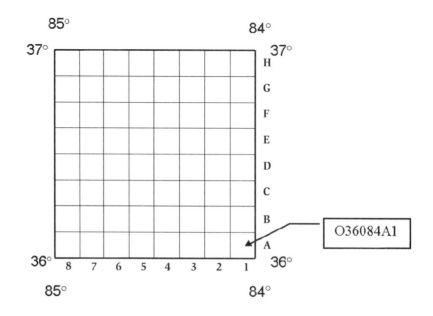

Fig. 2–3. Determining map index numbers.

# OBTAINING DRGS

In addition to the sources described in Chapter 1, the following data sources are available for DRGs.

## Earth Explorer

USGS runs a web site called Earth Explorer. It can be accessed at *http://earthexplorer.usgs.gov*. (A step-by-step example of using Earth Explorer can be seen in slides 10 through 17.) The Earth Explorer web site can be used to access many different USGS data sets. Some types of data are available free of charge, whereas others (such as DRGs) have a cost associated with them. Once you access Earth Explorer, you can choose the type of spatial data you wish to order and set the coordinates for the area you want to cover, as indicated in figure 2-4.

Slides 10–17

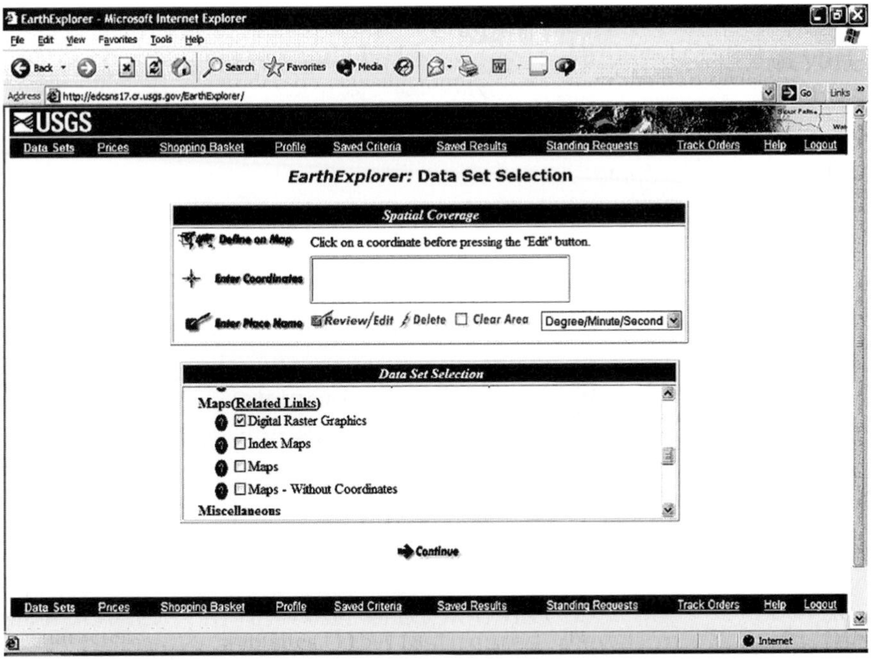

Fig. 2-4. Data selection options in Earth Explorer.

Once you have selected an area and type of spatial data, the Earth Explorer server looks for the available data. A list of data sets is displayed, and clicking on the list brings up a list of each DRG (or other spatial data) that meets your criteria. You can then choose to add whatever data sets you want to add to your shopping basket, as indicated in figure 2-5.

Fig. 2-5. Downloading DRG files in Earth Explorer.

Once you have completed the necessary purchasing information, the USGS will send you an e-mail that contains a hyperlink so that you can download your data.

## TerraServer

Slides 18–23

You can download DRGs in JPG format from TerraServer-USA, along with corresponding World files (JGW). (See slides 18 through 23.) To download these files, go to *http://terraserver-usa.com.* As you zoom in on an area, a corresponding DOQ (these are discussed in material to follow) or DRG will be displayed. You can then download the image for that area by clicking on the Download button on the web page. You can also retrieve the corresponding World file. Images from TerraServer were used to create the background layers of the MetroWeb mapping service at the Tennessee Electronic Atlas (*http://tnatlas.geog. utk.edu/tea/metrogis.htm*). Procedures for downloading from TerraServer can be found at *http://rockyweb. usgs.cr.gov/outreach/terraserver.html.*

# WORKING WITH DRGS

Most software that can display TIFF files can display DRGs. In addition, most users will likely want to use the DRGs with a GIS program, so that they can be georegistered against other data layers in the same coordinate space. DRGs are useful as a background layer for other data layers, for ground truthing data collected in the field, for plotting GPS points in some context, and for entering points from paper topographic maps into a GIS. There are several GIS packages for working with DRGs. The two discussed here are ArcGIS and Global Mapper.

ArcCatalog, a component of ArcGIS, will recognize all DRGs as raster image files. For those with metadata in FGD files (such as those you can get from Earth Explorer), ArcCatalog will recognize the metadata in the FGD file and create a metadata XML file containing all of the projection information. For other files, you will have to manually enter the projection information into ArcCatalog (see slide 24). If you do not define the projection information in Arc-Catalog, strange things can happen in ArcMap if the DRGs are in two different UTM zones. ArcMap will read TIFFs without projection information and will not adjust for the change in UTM zone. The result will be that

Slide 24

adjacent DRGs will be plotted far apart, possibly with the western-most image appearing far east of its neighbor.

Slides 25–28

In slides 25 through 28, there is an example using two TVA-generated DRGs: *O35090D1* and *O35089D8*. From the file names, you can tell that these are from 1:24000 topographic maps. The first contains the first quad in the D row of the area starting at longitude 90. The second contains the last quad in the D row of the area starting at longitude 89. The latitudes are the same (35 degrees, row D). Thus, the two maps should be adjacent. However, when displayed in ArcMap they are far apart and the eastern DRG is to the west of the western DRG. To fix this problem, you must define the projection details in ArcCatalog (see slides 29 through 31).

Slides 29–31

As long as the projection information is known, ArcMap will plot DRGs as images and place them properly on the data view. Once the image is loaded into ArcMap, you can use it as a background layer as you see fit. For example, you might want to combine the DRG with a DOQ or with vector data such as coordinates from GPS receivers.

The software Global Mapper can also be used with DRGs. If the program does not find a metadata file it recognizes (for example, an FGD file) or a projection file, it will try to guess the projection based on the file name. Figure 2-6 shows an example for a file named *O36084A2*. Realizing that the longitude is between 84 and 90 degrees, the program correctly assumes that the projection is UTM Zone 16.

Fig. 2-6. Global Mapper UTM zone dialog.

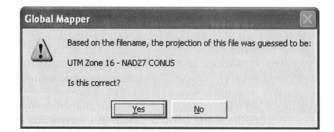

There are two ways to determine your UTM zone. One is to find a table of zones and their corresponding longitude ranges. Many GIS help files contain that information. The other way is to perform a simple calculation. UTM zones are 6 degrees wide from an eastern-most meridian (EM) to a western-most meridian (WM). The EM and WM are always divisible by 6. For zones in the western hemisphere, the relationship between the EM and the zone number is:

Zone = (180 - EM)/6

Suppose, for example, you were working in an area at 82 degrees west longitude. The eastern-most meridian evenly divisible by 6 is 78, and thus 78 is the EM of the zone. The calculation for zone number is:

Zone = (180 - 78)/6 = 17

The limits on the zone are from 78 to 84 degrees. Global Mapper has two very useful options for working with DRGs. It can clip the map collar from the image. Second, it can project the data into one of several projections, including State Plane or geographic data.

Although some data suppliers offer DRGs with the collar removed, you will usually have to use some software package to remove it. Accomplishing this in Global Mapper is as follows. After loading a DRG (or several DRGs) into Global Mapper, select Tools > Control Center. A dialog box will appear listing all DRGs (or other files) currently loaded into Global Mapper. Select the map you wish to clip and then click on Options. The dialog box shown in figure 2-7 will appear. Check the option Auto-Clip Collar to clip the parts of the image that are outside the map extent. If you wish, you can customize the clip area to leave only a portion of the map.

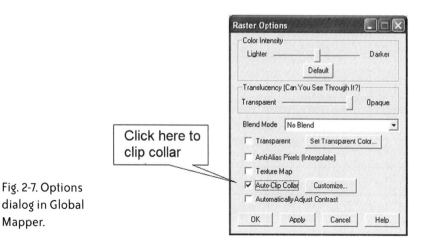

Click here to
clip collar

Fig. 2-7. Options
dialog in Global
Mapper.

To use a different projection, select Tools > Configure. Select the Projection tab to set the projection. Figure 2-8 shows the settings for reprojecting from UTM Zone 16 in meters with NAD27 as the datum to Tennessee State Plane in feet using NAD83 as the datum. Figure 2-9 shows the resulting map, which is the reprojected and clipped version of figure 2-1.

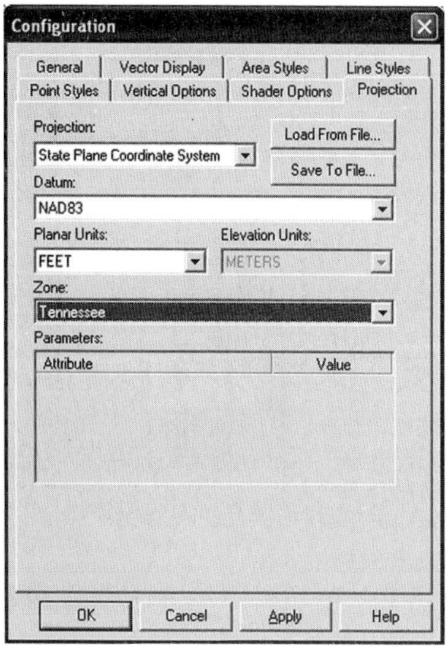

Fig. 2-8. Setting the projection in
Global Mapper.

Fig. 2-9. A clipped and projected map.

If you have many DRGs loaded into Global Mapper, you can export them to make sure they are all in the same projection.

# CROSSING UTM ZONES

At the 2003 Summer Assembly of the University Consortium for Geographic Information Science, Dr. Michael Goodchild quipped, "The most important point in any study will fall at the corner of four map sheets." To this clever comment, one might add that the sheets will fall in different UTM zones, and each sheet will be based on a different revision date!

Getting DRGs to edge match is not difficult as long as the DRGs you are working with are all based on the same projection, datum, and zone. A common problem occurs when crossing UTM zones. The maps seem to be placed in the wrong coordinate space. Consider two DRGs: *O35084A1* and *O35083A8*. From the file names, you can see that these DRGs represent adjacent areas.

The World file for O35084A1 lists the center of the northwestern-most pixel as 760900.05, 3891592.45. The World file for the adjacent file O35083A8 yields the coordinates of 225074.63, 3891646.8. If your GIS program knows only the information in the TIFF, it will place the second file too far west of the first file. Yet the DRGs should be adjacent, with O35083A8 to the east of O35083A1.

When combining DRGs, edge matching problems can occur. Discontinuities between adjacent DRGs are often obvious on maps consisting of two or more DRGs. Even when both DRGs are clipped and placed in a consistent map projection, there might be a discernible edge between them, as shown in figure 2-10.

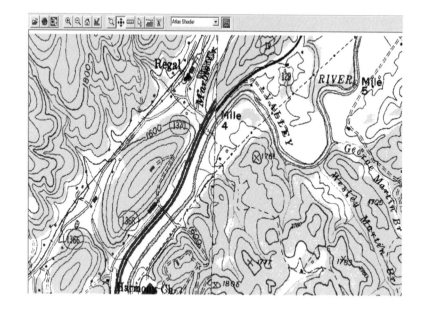

Fig. 2-10.
Adjacent DRGs
in the same
projection,
showing
discontinuities
at the map
edges.

# Digital Orthophoto Quads

Slide 32

The USGS distributes digitally rectified aerial photos in two file sizes. The 7.5-minute blocks are DOQs. The 3.75-minute blocks are called digital orthophoto quarter quads (DOQQs). (See slide 32.) DOQs are most commonly thought of as black-and-white images, but color-infrared and natural-color DOQs are also produced. You might wonder what is meant by the term *rectified*. When aerial photos are taken, the resulting picture is inherently distorted. That is, an aerial photo is not a map. The distortion comes from several sources. Local topography may distort our perception of the area. The recording mechanism (i.e., the camera) may be at an angle to the ground. The camera lens itself can cause distortion. The aircraft carrying the camera may not be parallel to the ground. Finally, not all areas of the photograph are directly beneath the camera. All such factors make raw aerial photography unsuitable for accurate mapping.

To adjust for these factors, the unrectified image is draped over a DEM of the area. Using information from the camera or other sensor device and the coordinates of known control points, the photo is then adjusted to register the photo, to remove displacement due to the factors cited previously, and to ensure that the ground control points are correctly positioned in the photograph. Photogrammetry is the science that creates a photograph that can be used as a map (i.e., a "rectified" map). Once the photo is rectified, it is resampled to generate a digital file. The resampling takes place every meter for DOQQs and every one or two meters for DOQs. Accuracy standards require that 90 percent of test points fall within 40 feet of the precise location for DOQs and 33.3 feet for DOQQs.

> **NOTE:** *For more technical details on rectification and other aspects of digital orthophotos, see Part 2: Standards for Digital Orthophotos available at* http://rockyweb.cr.usgs.gov/nmpstds/doqstds.html.

The USGS requires that all adjacent DOQs overlap, so that is possible to mosaic the photos without gaps. For DOQs produced under the National Digital Orthophoto Program, this overlap should be a minimum of 300 meters (see slide 33). DRGs, DOQs, and DOQQs can be displayed together. Those that represent the same area should align closely (see slide 34).

Slide 33–34

## Data Formats

DOQs and DOQQs come in four major formats. When ordered on CD, DOQs are in compressed JPG formats. However, the most widely used format is the GeoTIFF format, discussed previously. DOQ native formats are also available. This format consists of an ASCII header followed by binary information on the image. Images are also available in MrSID (Mulitresolution Seamless Image Database) format. This format achieves very high compression rates with little loss of information.

Both Global Mapper and ArcMap will read MrSID files directly. Global Mapper will also read native format DOQs and DOQQs.

*TIP: If your GIS software will not read native format DOQs, there is a free utility that converts them to Geo-TIFFs. That utility is available at* http://rockyweb. cr.usgs.gov/software. *The program name is* doq2 exe.

## Obtaining Digital Orthophotos

Most of the sites discussed as sources of DRGs also supply DOQs and DOQQs. In particular, you can order DOQs from the USGS through Earth Explorer (figure 2-5 shows an Earth Explorer shopping cart with one DOQ added), or otherwise obtain them from map libraries and public user groups, state GIS data portals, and the TerraServer-USA site. Some commercial sites also make them available, such as the GeoCommunity GIS Data Depot site (*www.gisdatadepot.com*). Some digital orthophotos may not be available or may have deleted sections for security reasons. As security concerns have grown since the September 11, 2001, terrorist attacks, spatial data for areas deemed sensitive have become less available. For a list of DOQ web sites, see slide 35.

Slide 35

## Working with DOQs

The information on ArcGIS and Global Mapper that applied to DRGs also applies to DOQs and DOQQs. As with DRGs, care must be taken when projecting these images or shapes of features can become distorted.

# WEB SITES AND DOCUMENTATION

There are several web sites and documents of interest to users of DRGs and DOQs. The most general descriptions of these data sets can be found at the Digital Backyard web site. The DRG web site is at *http://mapping.usgs. gov/ digitalbackyard/drgbkyd.html* (see figure 2-11), and the DOQ web site is *http://mapping.usgs.gov/digitalbackyard/ doqbkyd.html.* Both of these sites offer nontechnical overviews of their respective data sets.

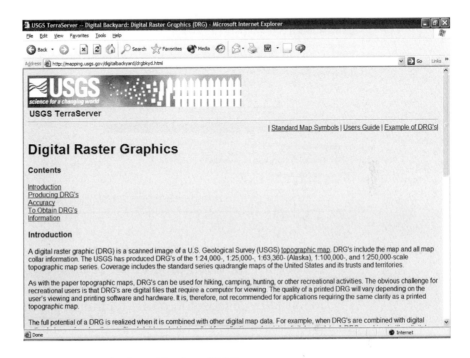

Fig. 2-11. The Digital Backyard DRG page.

Another overview of DRGs can be found at the Digital Raster Graphics Factsheet page at *http://erg.usgs.gov/ isb/pubs/factsheets/fs08801.html.* Detailed information on DRGs can be found via the DRG home page at *http:// topomaps.usgs.gov/drg.* The DRG home page explains how to obtain DRGs from USGS business partners, data partners (i.e., TVA and the California Spatial Information Library), and directly from USGS. There are also several links at the top of the page (see figure 2-12). Two links of interest are the Program Overview link and the Standards and Technical Documentation link.

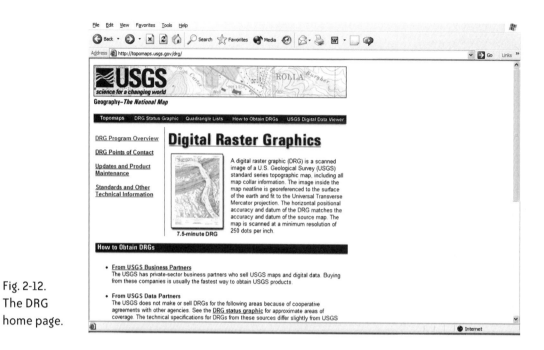

Fig. 2-12.
The DRG
home page.

The program overview web site contains detailed information about the DRG program. The page begins with a history of the DRG program and areas of coverage. This is followed by discussion of the technical details of DRG content. These technical details cover the following topics.

❏  Physical file formats and names

❏  Georeferencing procedures

❏  Data resolution

❏  Projections and datums

❏  Positional accuracy

❏  Colors

❏  Metadata

❏  Product distribution

Discussions of the content of the GeoTIFF, metadata, and World file follow. Included is a link to technical descriptions of the GeoTIFF format (see figure 2-13). A list of all available DRGs can be obtained from the Updates and Product Maintenance section of the DRG program overview page.

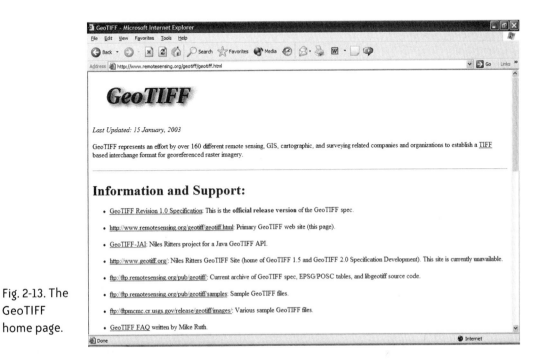

Fig. 2-13. The GeoTIFF home page.

DRG standards and technical documentation are available via links at *http://topomaps.usgs.gov/drg/drg_ technical. html*. Also available are technical papers on deciphering DRG file names, understanding Transverse Mercator projections, and on grasping the effect of datum shifts from NAD27 to NAD83. All of these standards documents and technical papers are available in PDF format. The paper on Transverse Mercator projections (*http:// topomaps.usgs.gov/drg/mercproj*), written by Larry Moore, contains a nice discussion of grids, graticules, and quadrangles. An understanding of these basics of projections and mapping is important when manipulating digital map products. As Moore points out, there is no best projection for digital data products.

As with DRGs, there exists a DOQ home page at *http://www-wmc.wr.usgs.gov/doq/* (see figure 2-14), fact sheet (*http://mac.usgs.gov/mac/isb/pubs/factsheets/fs05701.html*), and product page (*http://edc.usgs.gov/products/aerial/doq.html*). The product page contains sections on product descriptions, prices, sample data, and how to search for and order DOQs for specific areas (see figure 2-15).

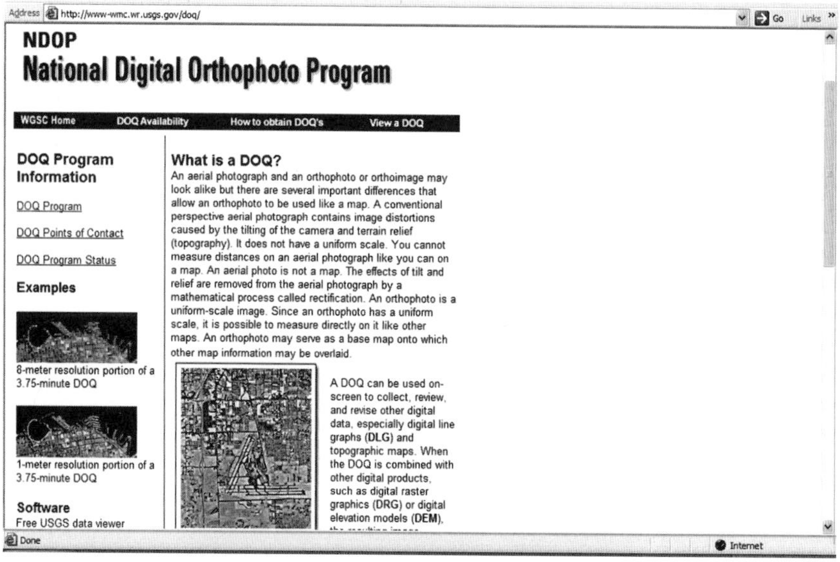

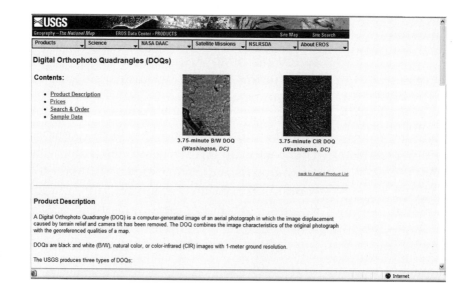

Fig. 2-14. The DOQ home page.

Fig. 2-15. The DOQ product page.

A description of the DOQ program can be found at *http://mapping.usgs.gov/www/ndop/*. Technical standards for DOQs, available as PDF files, can be accessed at *http://rmmcweb.cr.usgs.gov/public/nmpstds/doqstds.html.*

CHAPTER 3

# DIGITAL LINE GRAPHS

## INTRODUCTION

THE PREVIOUS CHAPTER DEALT WITH RASTER representations of topographic maps and the orthophotos from which such maps are derived. This chapter deals with the vector representation of topographic maps. Digital line graphs (DLGs) are vector representations of most, although not necessarily all, information found on topographic maps. Unlike some vector representations of map features (such as TIGER files, discussed in Chapter 6), features on a topographic map are separated into different DLG files according to feature type. DLGs separate information into themes, so that hydrographic features are in one DLG, transportation features in another, and so on. Table 3-1 presents an overview of DLGs.

**Table 3-1: Overview of Digital Line Graphs**

| Topic | Nature of DLG Representation |
|-------|------------------------------|
| Description | Vector representations of layers in topographic maps. |
| Format | Two formats are used: optional format and SDTS vector format. |

| Topic | Nature of DLG Representation |
|---|---|
| Scales | Three scales are used: 1:24000, 1:100000, and 1:2000000. |
| Projection/ coordinate system | Varies by area. Usually UTM in meters based on NAD 27 (called NAS in SDTS) or NAD83 (called NAX in SDTS). |
| Original source | USGS topographic maps. |
| Software considerations | Not all DLG translators support the one-to-many attributes. There are two major formats for DLGs that can dictate what tools you should use. |
| Other considerations | DLGs separate information on a topographic map into themes, such as hydrography, hypsography, transportation, and so on. DLGs serve as the basis for other data sets, such as NHD data (Chapter 8) and in some cases DEM data (Chapter 4). |
| Software used in this chapter | Global Mapper, DLG2SHP, ArcTools DLG Optional Format Import Wizard, ArcTools Vector SDTS Import Wizard. |

Slides 1–3

Topographic maps contain a wealth of information. To make the distribution of vector data more manageable, the USGS divides the information on topos into nine categories (see slides 1 through 3). These are as follows, along with the two-letter key found in the compressed (*gzip*) DLG file name.

❏  Hypsography (HP)

❏  Hydrography (HY)

❏  Vegetative Surface Cover (SC)

❏  Non-Vegetative Surface Cover (NV)

❏  Boundaries (BD)

❏  Survey Control Points and Markers (SM)

❏  Transportation (TR, or MT, RR, RD)

❐   Manmade Features (MS)

❐   U.S. Public Land Survey System (PL)

DLGs are currently distributed at three scales and in two formats. The three scales are 1:24000, 1:100000, and 1:2000000. The two formats are known as the Optional Format (sometimes called DLG-O) and the Spatial Data Transfer Standard (SDTS) vector format. A Standard Format (DLG-S) was discontinued in July of 1996. To make matters a bit more confusing, DLGs are often called DLG-3, whether they are in Optional or SDTS format. Finally, as topographic maps are updated, newer versions of DLGs are produced. It is easy to get confused between the newer and older versions. Unlike DRGs, you cannot completely determine the location and scale contained in a DLG by its file name.

There are some differences in how the two file formats (DLG-O and SDTS) organize information. Some SDTS files group all transportation features into one file. The DLG-O format and some SDTS files break transportation features into three groups. Group 1, designated with an RD in the file name, consists of roads and trails; group 2, designated with an RR in the file name, consists of railroads; and group 3 consists of pipelines, transmission lines, and miscellaneous transportation features (designated with an MT in the file name).

# DLG TOPOLOGY

DLGs are based on a topological data model. In this vector model, every line segment has a "from" node and a "to" node, and a left and a right polygon. Maps are modeled as a collection of points, lines, and polygons. A DLG, whether in SDTS or Optional format, will contain all three types of features. However, the two storage formats (DLG-O and SDTS) handle the topological features a bit differently.

The DLG-O format is most like the coverage model made popular by ESRI. In fact, the format of a DLG-O file is quite similar to an ESRI Ungenerate file. Thus, when working with DLG-O format files three types of topological entities can be generated: points, lines, and polygons. Figure 3-1 shows the polygons defined by roads in a DLG-O file, and figure 3-2 shows all lines (the roads themselves) defined by the same DLG.

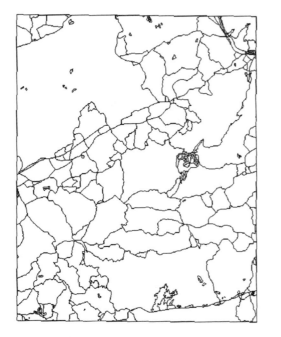

Fig. 3-1. Polygons defined by roads in a DLG-O file.

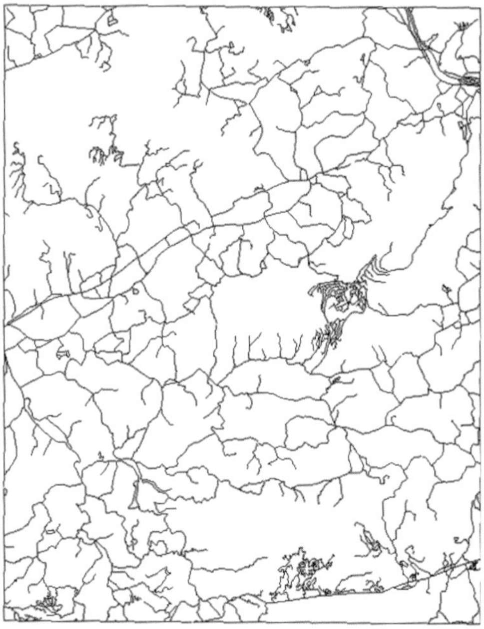

Fig. 3-2. Roads in the DLG-O file.

There are two things to note about these figures. In figure 3-1, not every road is represented. Roads that do not enclose an area, either by themselves or in combination with other roads, are not part of the polygon definition and are consequently not drawn in the figure. The line layer in figure 3-2 does contain every road in the DLG. The second thing to note is that boundary lines for the study area are in both figures. Thus, there is one more polygon present than you might realize. In particular, the entire study area is considered a separate polygon (sometimes referred to as the universe polygon), depicted in figure 3-3. This polygon is given an ID number of 1 in DLG-O. As will be pointed out in material to follow, Global Mapper and ArcTools filter out this polygon, whereas DLG2SHP does not.

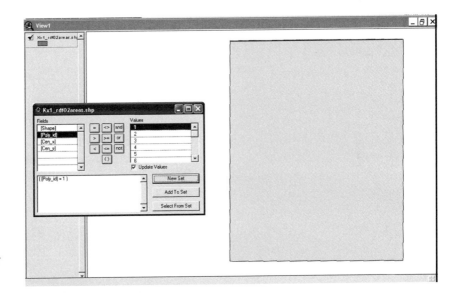

Fig. 3-3. Study area polygon for the DLG-O file.

1:2000000 DLGs present a slightly different case. Figure 3-4 shows what is known as the void area polygon. This is the area within the file rectangle but outside the study area. The void-area polygon will occur in both SDTS and DLG-O files at this scale. In addition, the DLG-O files will also contain the entire rectangle, similar to that shown in figure 3-3. Some 1:2000000 DLG-O files also contain degenerate lines. These are lines of zero length. Some GIS systems cannot render lines of zero length. At least one of the programs discussed in material to follow (DLG2SHP) generates points that correspond to the degenerate lines so that they may be rendered in a GIS as point features.

Fig. 3-4. Void area
polygon in
1:2000000 DLGs.

SDTS handles lines and polygons in a similar manner to
the Optional format. That is, all lines (called complete
chains in SDTS parlance) are considered a single type of
feature. The same is true of polygons (called GT-polygons
in SDTS). As with DLG-O files, there can be universe and
void polygons. However, SDTS files assign point features
to up to four classes: Area Points, Entity Points, Nodes,
and Generic Points. These classes are usually designated
with the identifiers NA, NE, NO, and NP, respectively (see
slides 4 through 8). Area points are similar to polygon
centroids (label points). There is one label point per polygon.

Slides 4–8

Entity points are spatial objects. For example, in a hypsography DLG there may be spot elevations. These would
be treated as NE or entity points. In hydrography DLGs,
springs or wells may be entity points. NOs (nodes) are
those points necessary to enforce planar topology. These
are equivalent to from nodes and to nodes in a coverage
model. Finally, the generic points (NPs) represent the registration points. For 1:24000 and 1:100000 DLGs these
correspond to the four corners of a quadrangle. At
1:2000000 there will usually be more registration points
than at just the corners.

Slide 9

Some SDTS DLGs will contain information on four 7.5-
minute quadrangles. If that is the case, a number is
assigned to each quad (see slide 9). For example, the
SDTS may contain area point layers named *NA01, NA02,
NA03,* and *NA04*—one set of area points for each quadrangle. Similarly, line features will have layer names of
*LE01, LE02, LE03,* and *LE04.*

# DLG ATTRIBUTES

The DRG and DOQ files discussed in Chapter 2 were raster files that had no separate attribute tables associated with them. There was merely a brightness value or a color value associated with each pixel. In contrast, vector data do have attributes. In particular, DLGs assign numeric codes to attributes, and the meaning of each code can be found in a lookup table. An important point to keep in mind is that not every feature has a single attribute associated with it. Features in DLGs, like many vector features, can have multiple attributes.

Each feature in a DLG may have zero, one, or many attributes. That is, there is a one-to-many relationship between features and attributes. Software vendors have taken two approaches to the problem of multiple attributes. For the one-to-may problem, some vendors allow up to five attributes, many of which will be blank or have a missing value flag. Others create additional attribute tables that allow for one-to-many relationships between map features and attribute values. Some vendors do not report features (e.g., polygons) if it is felt that those features have no substantive meaning. For example, polygons defined by roads may not be reported. The differing approaches used by three software vendors are discussed in material to follow.

Consider, for example, a major road. It might have a U.S. highway number, a state highway number, and a county highway number. It might also have a classification that indicates if it is divided, underpassing, or at the same location (coincident) with another feature, such as a fenceline, municipal boundary, or pipeline. In this case, that road segment will have separate attributes (corresponding to additional fields) for each highway number, road characteristic, and coincident feature.

Hypsographic contours present a particularly challenging case. There is a type of contour called a carrying contour. This is a line, found in steep areas, that has two elevations associated with it. This would be the case with a sheer rock face. In that case, the USGS assigns two contour values to the line: one for the upper elevation and one for the lower elevation.

Attributes are coded in DLGs, whether SDTS or Optional format, as numeric codes consisting of two parts: a three-digit major value and a four-digit minor value. The major value indicates the type of feature. For example, all major values that start with 18 (such as 180, 181, and 183) indicate features associated with railroads. Each major code will indicate if the minor value is to be interpreted as a code or as a numeric value. If the third digit of the major code is a zero, the minor value is a code, whose meaning can be found in a lookup table. Consider the entity code 1800610. The major code 180 indicates that the minor value (0610) is a coded value whose meaning can be found in a lookup table (see slide 10). In this case, the code is for a rapid-transit rail line.

Slide 10

A value of 1810002 is interpreted differently. The primary value of 181 indicates that the next four digits will contain the number of tracks present. In this example, there are two tracks. Fortunately, many translators interpret these values automatically and place their descriptive meanings in attribute tables, or supply lookup tables so that you can simply join the attribute codes to the features.

SDTS format files contain attribute modules. That is, attributes of similar type are grouped as a module. Some modules apply to all types of features. For example, every SDTS DLG will contain an AHDR attribute module. This module contains the header information, such as the map extent, vertical datum, and other attributes common to all DLGs. Each type of DLG may contain an ACOI module that lists coincident features, such as a stream that is also a political boundary line. Other modules may contain information specific to the type of theme represented in the DLG. Table 3-2 outlines the possible modules (these are in addition to the AHDR and ACOI modules). These various types of attribute modules lead to many possible relationships between map features and attribute tables.

## Table 3-2: SDTS Attribute Modules

| DLG Theme | Attribute Module Name | Description |
|---|---|---|
| Hypsography | **AHPF**, *AHPR*, and *AHPT* | Feature attributes, elevation in feet, elevation in meters |
| Hydrography | **AHYF** | Feature attributes |
| Vegetative surface cover | **ASCF** | Feature attributes |
| Non-vegetative features | **ANVF** | Feature attributes |
| Boundaries | **ABDF**, **ABDM**, BFPC (1:2000000 only), and BFPS (1:2000000 only) | Feature attributes, agency, county names, state name |
| Survey control markers | **ASMF** | Feature attributes |
| Transportation (MT, RD, and RR attribute files will be present in their respective file types: miscellaneous transport, road, or rail) | **AMTF**, **ARDF**, **ARDM**, **ARRF**, BMTA (1:2000000 only), and BFPS (1:2000000 only) | Feature attributes (miscellaneous), feature attributes (roads), route (roads), feature attributes (rails), airport names, state name |
| Manmade features | **AMSF**, BMSP (1:2000000 only), BFPC (1:2000000 only), BFPS (1:2000000 only) | Feature attributes, populated place names, county names, state name |

| DLG Theme | Attribute Module Name | Description |
|---|---|---|
| U.S. Public Land Survey System | **APLF** and BGRL (1:2000000 only) | Feature attributes, land grant names |

The items in bold in Table 3-2 correspond to SDTS vector profile primary attribute modules. These modules are required by the SDTS DLG standards. One of the modules shown in italics in Table 3-2 (*AHPR* or *AHPT*) must be present in a hypsography DLG.

# OBTAINING DLGS

As discussed in chapters 1 and 2, there are several places to obtain digital data. Those places that were discussed in the previous chapter will only be mentioned briefly here. In particular, you should consider map libraries, state user group web sites, USGS Earth Explorer, and USGS business partners.

> **NOTE:** *The current version of the download scripts for DLGs at Earth Explorer appends an extra .gz to the end of each downloaded DLG file name. (This may be fixed by the time you read this.) Thus, a file that should be named 1426779.HY.sdts.tar.gz ends up being named 1426779.HY.sdts.tar.gz.gz. This may cause problems for some unzipping products.*

Each DLG is *gzipped* and, if necessary, *tarred*. *tar* and *gzip* are utilities that were developed for UNIX systems. *gzip* is a file compression program. However, unlike PC-based compression utilities, *gzip* will compress only one file. For multiple files, a second utility, called *tar*, was developed to concatenate several files into one so that they could then be *gzipped*. SDTS DLGs consist of many files and thus come *tarred* and *gzipped* (see slide 11).

Slide 11

Optional format DLGs consist of only one file and thus are only *gzip*ped.

# USGS DATA DOWNLOAD

The good news about DLGs is that all scales and areas are available through free download sites. In addition to the Earth Explorer site, DLGs can be accessed via the USGS Data Download site (*http://edc.usgs.gov/geodata/*). You can choose to access the files available for download by graphics (point at a map), by state, or by alphabetical listing (see figure 3-5). For a detailed example of the steps required to download from the Data Download site, see slides 12 through 25. To obtain all DLGs for a particular state or area, you might want to contact a USGS business partner or a state GIS data supplier.

Slides 12–25

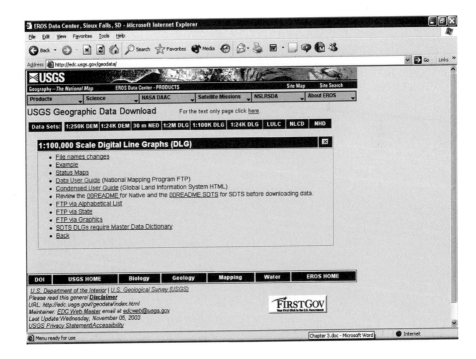

Fig. 3-5. USGS Data Download site.

# WORKING WITH DLGS

There are several software packages that will import DLGs into useful formats. The three discussed here are Global Mapper, DLG2SHP, and ArcToolbox. As discussed in Chapter 2, Global Mapper will automatically unzip files, and thus there is no need to un-*gzip* and un-*tar* the DLG files. Global Mapper has built-in rendering for the various types of information available in DLGs. For example, figure 3-6 shows an imported DLG road file.

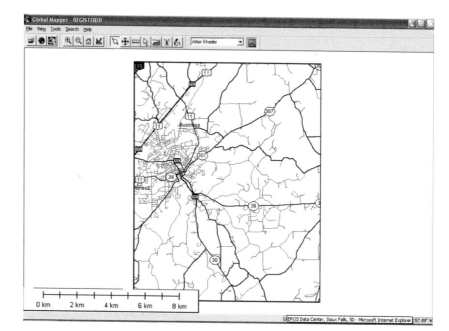

Fig. 3-6. A DLG
road file in
Global Mapper.

Slides 26–27

Note that Global Mapper does not display all of the point types mentioned previously. Only topological points that correspond to features with named attributes (such as dead ends, bridge abutments, and cul-de-sacs) are shown. When you export data from Global Mapper to a GIS format, such as an ESRI shapefile, the information not shown is not exported (see slides 26 and 27). That is, Global Mapper makes the decision as to what feature types and attributes are useful and filters all others out. This filtering takes place even if you specify a particular feature type. For example, the polygons defined by the roads in figure 3-6 are not exported to a shapefile by Global Mapper, even if you explicitly ask for them. Global Mapper's filtering logic makes working with DLGs relatively easy, but it also results in a loss of information.

Slides 28–31

DLG2SHP uses a different approach (see slides 28 through 31). The authors of DLG2SHP (this author and Dr. Cheng Liu) pass all of the information available in the DLG on to you. You can then use your own judgment in keeping or discarding the information you think is necessary. The program also calculates the lengths of all lines (in DLG units, usually meters) and reports all topology found in the DLG.

> **NOTE:** *The DLG2SHP User Manual and setup program can be found in the SOFTWARE directory of the companion CD-ROM.*

There are differences in the output files DLG2SHP generates for Optional format DLGs and SDTS format DLGs. For DLG-O files, the program creates point, line, and polygon shapes. These contain basic information about each type of feature.

Additional attribute files are also created. The files named *nodeatts*, *linkatts*, *areaatts*, and *dglnatts* contain attribute information for nodes, lines, areas, and degenerate lines, respectively. Further, if degenerate lines are present, equivalent node features for them are generated, along with their attributes. Figure 3-7 shows an Optional format DLG shapefile created by DLG2SHP, along with its line attributes. Note that for Optional format DLGs DLG2SHP automatically displays the definition of each code value. It will interpret the minor code value as a character string (Cvalue) or as an integer value (Ivalue), whichever is appropriate.

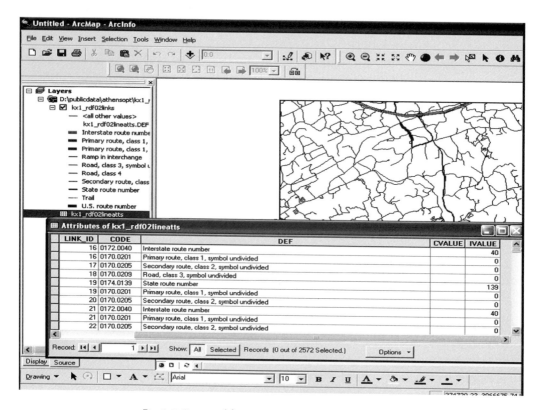

Fig. 3-7. Optional format DLG converted to shapes by DLG2SHP.

SDTS format DLGs classify points into four separate types: NA, NE, NO, and NP. DLG2SHP generates all four of these types, and thus potentially there can be up to six shape types generated from a single DLG (areas, lines, and the four types of nodes). As outlined in Table 3-2, SDTS files can contain several types of attribute files, one for each type of spatial feature.

The additional attribute files needed for these shapes are dependent on the type of feature being extracted. As discussed at the beginning of the chapter, DLGs assign spatial features to one of nine classes, and the SDTS transfer standards contain modules (outlined in Table 3-2) for each class of feature. DLG2SHP creates a corresponding attribute table for each module-feature type combination (see slides 32 through 34). For example, an attribute file name of *LE02_HY01AHYF.dbf* will contain feature attributes for the second set of line features in a hydrography SDTS format DLG. *PC02_HY01AYF* will contain attributes (such as lake names) for polygon features for the second 7.5-minute quad area in the SDTS DLG. In the case of hypsography, DLG2SHP will automatically assign elevations to the resulting line and point shapes. Figure 3-8 shows such a shape rendered in three dimensions.

Slides 32–34

Fig. 3-8.
Hypsography DLG
rendered in 3D.

DLG files often have file names that provide much information to the data user. For SDTS files, DLG2SHP gives the directory that holds the resulting shapes a descriptive name consisting of the area name, layer type, and the month, day, hour, minute, and second the program finished execution. An example of such a directory name is *CHATTANOOGA_TN_HYDROGRAPHY1_23_14_05_02*. This would be the hydrography layer for Chattanooga, Tennessee, which would have been created on January 23 at 2:05:02 PM.

The program DGL2SHP does not place descriptive names in the attribute fields when using SDTS-format DLG files. However, it does supply a lookup table (*entity.dbf*) so that you can join the attribute descriptions to their corresponding codes. Details are presented in the corresponding PowerPoint file. SDTS files are accompanied by metadata about their content. This information is reported by DLG2SHP in the output directory in a "*.doc*" file and a "*.log*" file.

DLGs can also be imported using ArcTools. However, it is necessary to first un-*gzip* and, if necessary, un-*tar* the DLG.

> **TIP:** *When using WinZip to un-tar files, note that Win-Zip has an option (usually under the Configurations menu) called Smart TAR CR/LF. The default configuration of WinZip leaves this option on. You must turn this option off before using WinZip to un-tar a file. If you do not, the resulting files will not be SDTS format compliant.*

Slides 35–38

When Optional format DLGs are un-gzipped, they produce single files that usually have a file extension of either .opt or .dlg. ArcTools contains a module for creating coverages from these files (see slides 35 through 38). Figure 3-9 shows ArcTools' DLG to Coverage tool in action.

Slides 39–42

For SDTS format DLGs, ArcTools contains an SDTS vector translator (see slides 39 through 42). The key to understanding this wizard is to know how un-*tarred* SDTS files are named. Each SDTS archive names files with a four-character prefix followed by the SDTS module name. For example, a catalog module for an SDTS file might be named *HY01CATD.DDF*. Every file in that SDTS set would start with the prefix *HY01*. You should be careful when un-*gzip*ping and un-*tarr*ing SDTS files. Every hydrologic SDTS file will have files that start with *HY01*. It is very easy to overwrite existing SDTS modules when un-*gzip*ping them. To be safe, each un-*gzip*ped and un-*tar*red archive should be placed in its own directory.

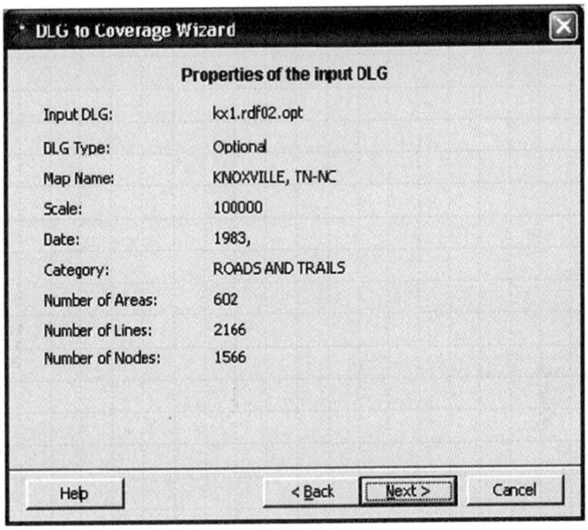

Fig. 3-9. ArcTools Optional format DLG Import dialog.

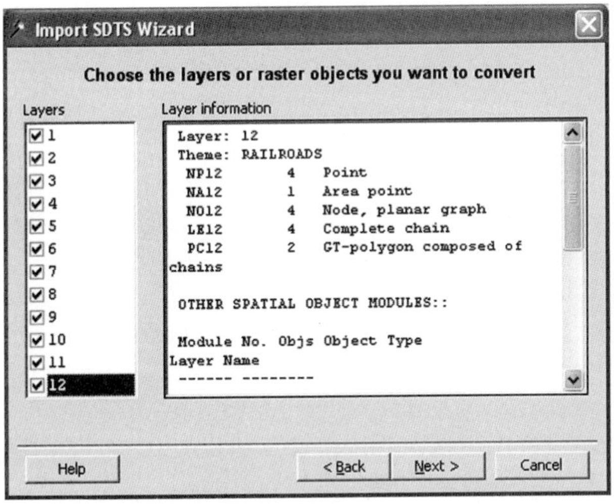

Fig. 3-10. ArcTools' SDTS vector format Import wizard.

The SDTS import wizard in ArcToolbox allows you to choose those types of features you wish to extract (see figure 3-10). You can then name the corresponding coverages you wish to create. The coverage names should consist of no more than 11 characters, and should not contain spaces or begin with a numeral. The SDTS import wizard will create Info files for each module listed in Table 3-2 that is present in the DLG archive.

# COMPARING **DLG** IMPORT **RESULTS**

The three methods discussed here present the user with a continuum of options for importing DLGs. Global Mapper makes decisions on what to keep and what to ignore for the user. DLG2SHP makes no such decisions and reports all information in the DLG to user. The SDTS vector file Import wizard in ArcToolbox takes a middle approach. It allows you to choose those coverages you wish to create. However, unlike the first two methods, it does not automatically un-*gzip* and un-*tar* the DLG archives. You must do that. Further, it requires that you understand the relationships between the coverages it creates and all the Info files it generates.

For Optional format DLGs, both Global Mapper and Arc-Tools allocate five fields for major and minor codes of each entity. In addition, ArcTools adds line length to line fields. DLG2SHP also reports line length. The one-to-many issue (one feature may have many attributes) is handled differently in DLG2SHP than in the other two programs. The entity codes and their interpretations are placed in a separate file for each feature type (points, lines, and polygons), which can then be linked or joined to the features (see slide 43).

Slide 43

Global Mapper will not report feature types (for example, points or polygons) that do not have attributes. Figure 3-11 illustrates the attribute tables for road lines generated by Global Mapper, DLG2SHP, and ArcTools for Optional format DLGs. All three approaches report the same information. However, Global Mapper and ArcTools generate many empty or missing value entries. DLG2SHP requires that you link the attribute table to the features based on the *Link_ID*.

For SDTS format DLGs, DLG2SHP generates separate attribute files for each type of feature, including all four node types (assuming they are present). It will automatically add elevations to hypsographic DLGs. Global Mapper filters out those node types and polygons it feels are unnecessary. ArcTools' SDTS vector import function will create an Info file for each module type listed in Table 3-2 that is present in the SDTS file.

**Global Mapper allocates 5 fields for major and minor line attributes**

**Attributes of gmexprt**

| LAYER | DLGMAJ_0 | DLGMIN_0 | DLGMAJ_1 | DLGMIN_1 | DLGMAJ_2 | DLGMIN_2 | DLGMAJ_3 | DLGMIN_3 | DLGMAJ_4 | DLGMIN_4 |
|---|---|---|---|---|---|---|---|---|---|---|
| ROAD OR STREET, CLASS 3 | 170 | 209 | 0 | 0 | 0 | 0 | 0 | 0 | 0 | 0 |
| PRIMARY ROUTE, CLASS 1, SYMBOL UNDIVIDED | 170 | 201 | 174 | 139 | 0 | 0 | 0 | 0 | 0 | 0 |
| PRIMARY ROUTE, CLASS 1, SYMBOL UNDIVIDED | 174 | 139 | 170 | 201 | 0 | 0 | 0 | 0 | 0 | 0 |

Record: 1  Show: All Selected  Records (0 out of 2166 Selected.)  Options

**ArcGIS does the same**

**Attributes of athensln arc**

| LPOLY# | RPOLY# | LENGTH | ATHENSLN# | ATHENSLN-ID | MAJOR1 | MINOR1 | MAJOR2 | MINOR2 | MAJOR3 | MINOR3 | MAJOR4 | MINOR4 | MAJOR5 | MINOR5 |
|---|---|---|---|---|---|---|---|---|---|---|---|---|---|---|
| 79 | 78 | 692.870667 | 301 | 491 | 170 | 209 | -99999 | -99999 | -99999 | -99999 | -99999 | -99999 | -99999 | -99999 |
| 78 | 88 | 928.567932 | 300 | 492 | 170 | 209 | -99999 | -99999 | -99999 | -99999 | -99999 | -99999 | -99999 | -99999 |
| 82 | 84 | 736.126892 | 296 | 493 | 170 | 201 | 174 | 139 | -99999 | -99999 | -99999 | -99999 | -99999 | -99999 |

Record: 0  Show: All Selected  Records (0 out of 2166 Selected.)  Options

**Attributes of kx1_rdf02links**

| LINK_ID | FROMNOD | TONODE | LPOLY | RPOLY | LENGTH |
|---|---|---|---|---|---|
| 491 | 374 | 365 | 143 | 140 | 692.9855 |
| 492 | 374 | 364 | 140 | 147 | 928.4891 |
| 493 | 375 | 360 | 144 | 150 | 736.0811 |

Record: 1  Show: All Selected  Records (0 out of 2166 Selecte

**Attributes of kx1_rdf02lineatts**

| LINK_ID | CODE | DEF | CVALUE | IVALUE |
|---|---|---|---|---|
| 491 | 0170.0209 | Road, class 3, symbol undivided | | 0 |
| 492 | 0170.0209 | Road, class 3, symbol undivided | | 0 |

ds (0 out of 2572 Selected.)

**DLG2SHP reports all attributes in a file that can be joined to the shape by line_id**

267390.70 3961714.11 Unknow

Fig. 3-11. Comparing Optional format attribute tables.

The method you use to import DLGs into shapes or coverages is, to a degree, a matter of taste. Global Mapper is easy to use and can clip and project files. However, it does filter out data you may want. DLG2SHP also is easy to use, but you will have to link attribute tables to feature attribute tables. The only exception is for hypsography, where elevations are automatically added to the resulting shapefile. ArcTools creates coverages for both optional and SDTS-format DLGs. The Optional format Import

Slide 44

wizard is fairly easy to use. The SDTS format option uses a generic SDTS vector format import function. It is a bit more difficult to determine how the resulting Info files relate to the coverages generated. However, a relate table, similar to one created by a *RELATE SAVE* command in Workstation ArcInfo, is created (see slide 44).

# WEB SITES AND DOCUMENTATION

As with DRGs and DOQs, the USGS provides a product page (*http://edc.usgs.gov/products/map/dlg.html*), user guide (*http://edc.usgs.gov/guides/dlg.html*), and fact sheet (*http://mac.usgs.gov/mac/isb/pubs/factsheets/ fs07896t.html*) for DLGs. The product page contains sections on product description, pricing, how to search for and order DLGs, and links to sample data.

The user guide consists of more detailed information on the background and coverage of DLGs (*http://edc. usgs.gov/guides/dlg.html*). The coverage information pertains to the three levels of the data set: large scale (greater than 1:100000), intermediate scale (1:100000), and small scale (1:2000000). Documentation on the content, including data attributes, can be found at the DLG standards page (*http://rockyweb.cr.usgs.gov/nmpstds/ dlgstds.html*) .

DLG standards documentation is divided into three parts. Part 1, General, describes the DLG program, data sources, DLG layers, and topological structure used. Part 2, Specifications, covers topics such as coordinate systems and map projections used, data quality, and errors. The attribute codes are described in Part 3, Attribute Coding. Features that are no longer represented in DLGs are simply crossed off the list of collected attributes (see figure 3-12). The attribute codes listed in this section of the document are also available in the *entity.dbf* file of the companion CD-ROM to this book.

```
Standards for Digital Line Graphs
Part 3:  Attribute Coding

         o  Single-point attribute codes

            None

         o  General purpose attribute codes

                170 0401  Traffic circle
                170 0402  Ramp in interchange
                170 0403  Tollgate
                170 0404  Weigh station
                170 0405  Nonstandard section of road
                170 0406  Covered bridge

         o  Descriptive attribute codes

                170 0600  Historical
                170 0601  In tunnel
                170 0602  Overpassing, on bridge (except drawbridge)
                170 0603  Under construction
                170 0604  Under construction, classification unknown
                170 0605  Labeled "Old Railroad Grade"
                170 0606  Submerged or in ford
                170 0607  Underpassing
                170 0608  Limited access
                170 0609  Toll
                170 0610  Privately operated or restricted use
                170 0611  Proposed
                170 0612  Double-decked
                170 0613  In-service facility, rest area, or roadside park
                170 0614  Elevated
                170 0615  Bypass
```

Fig. 3-12. Road attribute codes for DLGs (source: "Standards for Digital Line Graphs, Part 3, Attribute Coding," USGS, 1998).

As mentioned previously, DLGs are usually distributed in one of two format categories: Optional or SDTS. The SDTS was developed to facilitate the transfer of spatial data between various computer systems without a loss of information. There are standards for several types of spatial data, including raster data, transportation networks, points, CADD drawings, and vector data. Each of these types is implemented through a profile that provides rules for storing specific object types, data requirements, and module naming. For DLGs, the appropriate profile is the topological vector profile (TVP). There is online documentation of the TVP content and standards (see figure 3-13).

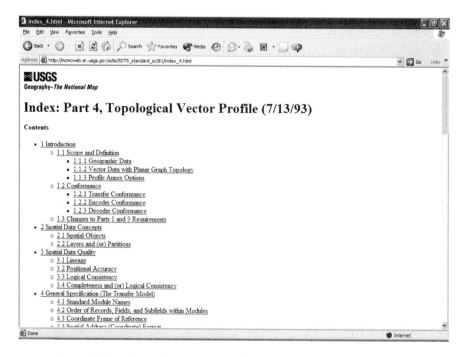

Fig. 3-13. The TVP online documentation.

The TVP is highly structured and any software designed to work data in TVP format will necessarily consist of modules to manipulate that structure. Fortunately, there is a public domain library of C++ objects for programmers to use when working with SDTS files. Developed by programmers Mark Coletti, Jamie Moyers, and Dave Edwards at the USGS Mid-Continent Mapping Center, the library consists of source code for reading and writing SDTS files (see figure 3-14). This is a valuable resource for GIS programmers who wish to work with SDTS format files (see slide 47).

Slide 47

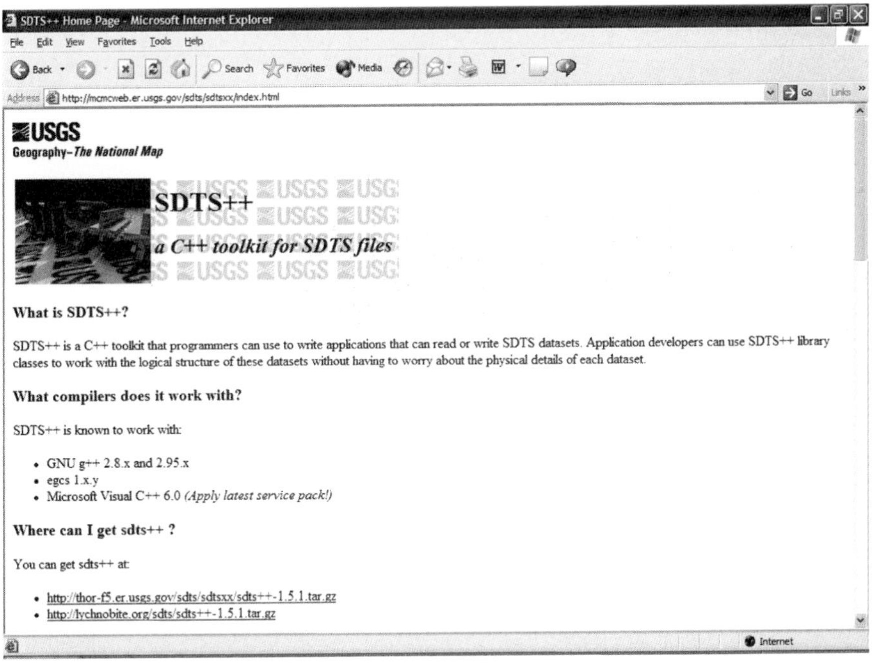

Fig. 3-14. The SDTS++ home page.

CHAPTER *4*

# DIGITAL ELEVATION DATA

## INTRODUCTION

ELEVATION DATA ARE USED FOR A VARIETY of purposes, including giving shading and relief characteristics to orthophoto or DRG overlays, conducting line-of-sight studies, producing view sheds, determining watersheds and water courses, conducting wildlife management, and locating cell phone towers. Their uses in national security and military applications are growing.

Slides 1–5

In this chapter you will learn about three sources of elevation data: digital elevation models (DEMs), the National Elevation Dataset (NED), and elevation data from the Shuttle Radar Topography Mission (SRTM; see slides 1 through 5). In the PowerPoint slides for Chapter 1 (slide 13) you saw how DLGs (digital line graphs) and DRGs (digital raster graphs) of the same place and scale register to the same coordinate space. This is because the primary source of data for creating and updating these data

71

sources is the National Aerial Photography Program (NAPP). NAPP air photos are the sources from which the data sets so far discussed are derived. The same will be true with DEMs; they will closely match a corresponding DRG or hypsography DLG.

As with DRGs and DLGs, DEMs will usually be in the coordinate system of the corresponding topographic map. For much of the United States, this means that the data will be in the Universal Transverse Mercator (UTM) coordinate system. For each quadrangle, the DEM reports X, Y, and Z values, usually in meters. As with DRGs, DOQs (digital orthophoto quads), and DLGs, problems can arise when crossing UTM zones of DEMs and in edge matching. As you might expect, there are some issues of accuracy, error, and indeterminacy (locations for which elevation cannot be determined) to be considered. The 1:24000-scale, 7.5-minute DEMs consist of elevations spaced either 10 or 30 meters apart.

A more recent (1999) set of elevation data, the NED, offers a seamless set of elevation data for the United States in decimal degrees. Elevations are assigned to points at every one arc-second (approximately 30 meters). This data set has been derived from 1:24000 DEMs and 1:63360 Alaska DEMs. The data from the DEMs were further processed to address some sources of error (discussed in material to follow) that sometimes crop up in DEMs.

However, many 7.5-minute DEMs have 10-meter elevation spacing. Therefore, some resampling has been used to generate the 30-meter NED data. NED data is available via the interactive web site *http://seamless.usgs.gov* (see figure 4-1). This makes it much easier for researchers who wish to use elevation data over a large area to obtain it. They simply download their area of interest. The NED data set they receive has consistent resolution, horizontal and vertical datums, and units of elevation.

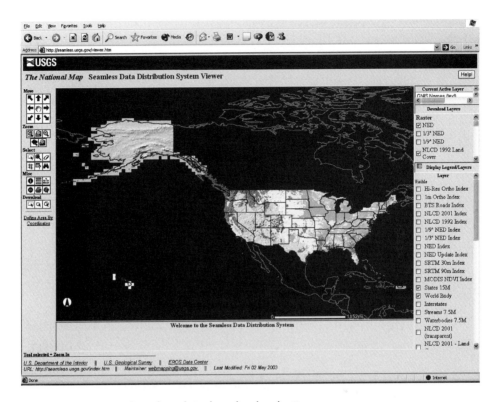

Fig. 4-1. Seamless data download web site.

Slides 6–7

For users requiring 30-meter resolution, using NED data is a fairly simple task. For those requiring 10-meter resolution, there may be NED data available at that resolution (see slides 6 and 7). If not, original 10-meter DEMs (which will likely be in the UTM coordinate system, if available) may be used. A major difference between NED and DEM data resides in the units used to measure location and elevation. DEMs have consistent units (usually meters) for X, Y, and Z locations. However, the "Z" values in NED are not in the same units as the "X,Y" locations. You will need to account for this difference when computing properties such as slope, aspect, and hillshading.

A more recent and extensive source of data is the SRTM. This data will cover 80 percent of the earth's land surface between the parallels of 56° South and 60° North. The data is output at a coarser scale than that for DEMs and NEDs, but SRTM data is much finer than that which was previously available for much of the world (i.e., 1 km). One can use SRTM data to generate an accurate global scale topographic map. Currently, data for much of North and South America are available. Table 4-1 presents an overview of these three data sources.

**Table 4-1: Overview of Elevation Data Sources**

| Item | Digital Elevation Models | National Elevation Dataset | Shuttle Radar Topography Mission |
|---|---|---|---|
| Description | Elevation by location, usually in UTM coordinates. | DEMs projected and mosaicked to produce a seamless national elevation data set. | Data collected by the Space Shuttle to produce a global topographic data set. |
| Format | Available in a DEM native format that most GIS programs can import, and in SDTS raster format. | Distributed in ArcGrid, BIL, TIFF, and GridFloat formats. | Same as NED. |
| Scales | Depends on source maps. Usually match 7.5-minute quads, 15-minute quads, and 1-degree DEMs. For 7.5-minute quads, elevation points are spaced every 10 meters or every 30 meters. | Same as underlying DEM. Some areas are based on 10-meter DEMs (before projection), others on 30-meter DEMs. | 30-meter and 90-meter spacing for much of the United States; 90-meter spacing outside the United States. As of January 2004, North and South America are the only available areas outside the United States. |

| Item | Digital Elevation Models | National Elevation Dataset | Shuttle Radar Topography Mission |
|---|---|---|---|
| Projection/ coordinate system | Varies by area. Usually UTM in meters based on NAD 27. X, Y, and Z are in the same units (meters). | North American Datum of 1983. The latitude and longitude are in decimal degrees, whereas the elevations are in meters. | World Geographic System 84 Datum. The latitude and longitude are in decimal degrees, whereas the elevations are in meters. |
| Vertical datum | National Geodetic Vertical Datum, 1929 (NGVD29). | North American Vertical Datum 88 (NAVD88) except for Alaska, which is NGVD29. | NAVD88. |
| Original source | Dependent on level. | Aerial photography and hypsographic DLGs are major sources of 10- and 30-meter DEMs. | Radar interferometry from shuttle missions. |
| Software considerations | Many GIS programs can read DEM format. | TIFFs can be read by many viewers. Many GIS programs can read BIL, BIP, ArcGrid, and GridFloat formats. | Same as for NEDs. |
| Other considerations | Care must be taken when projecting or combining across UTM zones; edges of adjacent areas may not match. | Care must be taken when hillshading, as X, Y, and Z coordinates are not in the same system. | Same as for NEDs. |
| Software used | Global Mapper, ArcTools, ArcMap, ArcScene. | Global Mapper, ArcTools, ArcMap, ArcScene. | Global Mapper, ArcTools, ArcMap, ArcScene. |

Elevation data is used for a variety of purposes. In instances where you wish to combine elevation data with other data sources (such as DOQQs, DLG layers, or land cover data), it is probably best to use the elevation source that uses the same datum and projection (if projected). If you need to use elevation data across UTM zones, NED data is useful. However, you will have to transform data from other sources to have the same datum and units as NED (discussed in material to follow). For areas outside the United States, SRTM data gives selected but growing international coverage.

# DIGITAL ELEVATION MODELS

DEMs are a matrix of elevation values wherein the rows and columns of the matrix correspond to locations on a map. The elevations are sampled at a specific horizontal spacing for each type of DEM. Figure 4-2 shows the sampling scheme for a 30-meter DEM based on a UTM area. Note that the collection of data points does not define a rectangle. It defines a quadrilateral. This is because the DEM points reflect the extent of a topographic map. Each filled dot in the figure will have an elevation level. For DEMs based on the UTM coordinate system, such as from hypsographic DLGs, it is possible to have evenly spaced data points. That is, each data point is separated by a fixed and symmetric distance, usually 10 or 30 meters, from every other data point.

For DEMs based on sampling done as units of spherical coordinates (such as latitude/longitude or the geographic coordinate system), the spacing between points varies, depending on the location of the data on the earth's surface. That is, these DEMs are based on spherical coordinates (arc seconds), not planar coordinates (meters). The 7.5- and 15-minute DEMs for Alaska, 30-minute DEMs, and 1-degree DEMs sample elevations at a spacing based on spherical coordinates, not fixed ground units. Spherical coordinate spacings are given in arc seconds, or fractions thereof. An arc second is 1/3,600 of a degree of latitude or longitude. Because the distance between lines of longitude varies on the earth's surface, the spacing between points is not fixed in all directions for such DEMs.

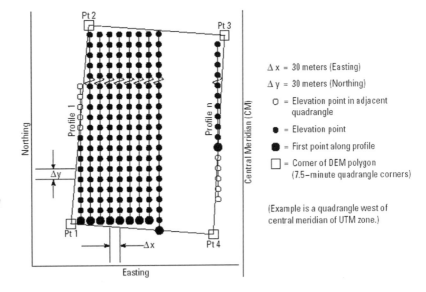

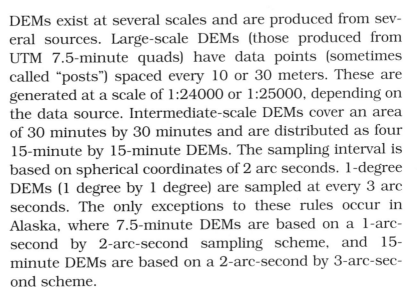

Fig. 4-2. Sampling scheme for 30-meter DEM (source: *Standards for Digital Elevation Models*, U.S. Department of Interior).

DEMs exist at several scales and are produced from several sources. Large-scale DEMs (those produced from UTM 7.5-minute quads) have data points (sometimes called "posts") spaced every 10 or 30 meters. These are generated at a scale of 1:24000 or 1:25000, depending on the data source. Intermediate-scale DEMs cover an area of 30 minutes by 30 minutes and are distributed as four 15-minute by 15-minute DEMs. The sampling interval is based on spherical coordinates of 2 arc seconds. 1-degree DEMs (1 degree by 1 degree) are sampled at every 3 arc seconds. The only exceptions to these rules occur in Alaska, where 7.5-minute DEMs are based on a 1-arc-second by 2-arc-second sampling scheme, and 15-minute DEMs are based on a 2-arc-second by 3-arc-second scheme.

You might wonder why different sampling schemes exist for Alaska. This is because near the poles the distance represented by a unit of longitude, such as 1 degree, is much shorter than the distance between a unit of latitude. Assuming a relatively spherical Earth, 1 degree of latitude has the same length everywhere, but the same cannot be said about 1 degree of longitude. This is because geographic coordinates are not rectangular. Put another way, the width of 1 degree of longitude at the equator is not the same as 1 degree of longitude at the pole. At the equator the width of a degree of longitude is about 69.17 miles, whereas at the poles it is 0 miles. This has real implications for data with horizontal sampling based on "arc seconds." Consider, for example, the data set downloaded from the NED data corresponding to the western part of Knox County, Tennessee. According to the metadata for this elevation data, the cell spacing of grids is 0.000278 (see slides 8 through 11). This is the number of degrees in 1 second. Put another way, 0.000278 = 1/3600. One degree of latitude is approximately 111312.27 meters. If each data point in the $y$ axis is 1 arc second, it is 0.000278 * 111312.27, or 30.92 meters.

Slides 8–11

At any latitude, the length of a degree of longitude must be adjusted by the cosine of the latitude to obtain its length. Thus, for the grid in question (44046839), the latitude is approximate 36° North. The cosine of 36° is 0.809. Thus, 1 degree of longitude is approximately 90053.52 meters. Converted to seconds, this yields 25.01 meters per second. Thus, for data based on arc seconds, the distance between sample points in the north-south direction will always be greater than or equal to the distance between points in the east-west direction.

DEMs are classified into three levels, depending on the source material used. The levels reflect how information was processed to create the DEM. A Level 1 DEM is based on aerial photography. Level 2 DEMs are based on hypso-graphic or hydrographic data sources, either photogram-metrically or from existing maps of up to 1:100000 scale. Level 3 DEMs are based on several DLG layers that have been integrated, taking into account features such as hypsography, hydrography, ridge lines, and other rele-vant features. Each of these levels has its own vertical accuracy standard.

## Accuracy

Estimates of elevation are a function of the vertical datum used when modeling the shape of the earth. In Chapter 1, you read about the various horizontal datums (such as NAD27, NAD83, and WGS84) used to estimate locations on the earth. These are used to estimate hori-zontal locations. When determining elevations, a vertical datum must be used. For DEMs, this is the National Geo-detic Vertical Datum of 1929, or NGVD29. This datum is based on mean sea level estimates from 26 tide gauges spaced along the east and west coasts of North America. Elevations are estimated, usually in meters but some-times in feet or decimeters, relative to NGVD29. Hawaii and Puerto Rico use locally defined mean sea levels for determining elevation.

It is assumed that all X-Y locations in a DEM are accu-rate. What may be inaccurate are the Z values, or eleva-tions. The vertical accuracy is measured by the vertical root mean square (RMSE) error statistic. This is calcu-lated as

$$RMSE = \sqrt{\frac{\sum (Z_i - Z_{true})^2}{n}}$$

where $Z_i$ is the DEM estimated elevation, $Z_{true}$ is the known true elevation, and $n$ is the number of sample points. The RMSE is calculated using a minimum of 28 test points, with at least eight of these being on the boundary of the area. For Level 1 DEMs, an RMSE of less than or equal to 7 meters is the desired accuracy. A value of 15 meters is the most allowed. For Level 2 DEMs, the desired accuracy is less than or equal to one half the original contour interval of the source map (usually a hypsographic DLG), with the full contour interval being the maximum allowable error. For Level 3 DEMs, the desired accuracy level is less than or equal to 1/3 of the contour interval, with 2/3 of the contour interval being the maximum allowed. (Source: *http://edc.usgs.gov/guides/dem. html#accuracy.*)

Errors in DEMs are classified into one of three groups: blunders, systematic errors, and random errors. Blunders are due to things such as miscoding, misreading contours from a DLG, or some other mistake. They are usually easy to spot. Systematic errors can result from the way data are collected. Scanning contours can lead to a phenomenon known as striping, where systematic errors are created in parallel bands. This reflects the parallel passing of the scanner over a map of hypsography. This is particularly a problem in relatively flat areas. Random errors cannot be predicted, but they do occur. For example, sinks (depressions, or low areas on a surface) that should not be present may occur in the DEM. The presence of sinks can cause hydrologic models, a major application of DEMs, to yield erroneous results.

In addition to errors, there are two other types of problems of which you should be aware. The first of these is void areas. As discussed in Chapter 3, DLGs can have void areas. For hypsographic DLGs, these correspond to areas where contours are unknown (perhaps due to features blocking contours) or to areas where contours are dropped due to mining or other disruptive activity. Elevations can also be disrupted in suspect areas. These correspond to disruptions of the surface. Disruptions can occur for many reasons, including lava flows, construction cut-and-fills, and landfills. In void areas, a flag is set and a value –32767 is set for the elevation. In suspect areas, an estimate of the elevation (the "presumed elevation") is used.

## Edge Matching

Even when in the same zone, UTM-based DEMs will not edge match perfectly. This is because their map extents do not exactly match. For example, the DEM Nashville West has as its UTM SE coordinates 522465.562500, 3997630.500000. The "adjoining" DEM (Nashville East) has as its SW coordinates 522446.125000, 3997636.750000. When plotted, there is a small gap between the two DEMs at the southern part of the map, as shown in figure 4-3.

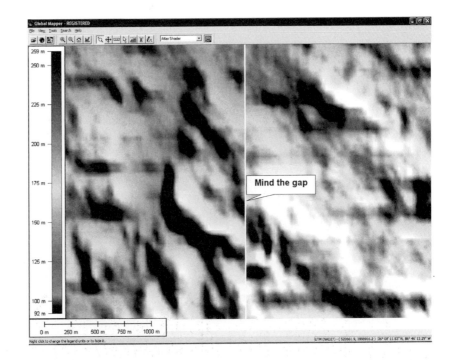

Fig. 4-3. Gaps between neighboring DEMs.

You may recall from Chapter 1 that the UTM projection is conformal, so that distances are only true along the central meridian. At the edges, distances are subject to a scale factor of 0.9996. As a result, the horizontal distance between post locations at the southern edges of adjacent DEM quadrangles may exceed the regularly spaced distances within a quadrangle. The result is a gap.

Perhaps even more troublesome is that areas that appear to overlap do not always have the same elevations, an example of which is shown in figure 4-4. The reason for the "overlap" is similar to that for gaps. That is, the distance between posts in the northern sections of adjacent DEM quadrangles may be less than regularly spaced distances within a quadrangle. Thus, the posts appear to overlap, and their elevations may be different. The software packages discussed in the sections that follow contain algorithms for extrapolating to fill gaps and for smoothing data between adjacent DEMs with non-matching overlaps.

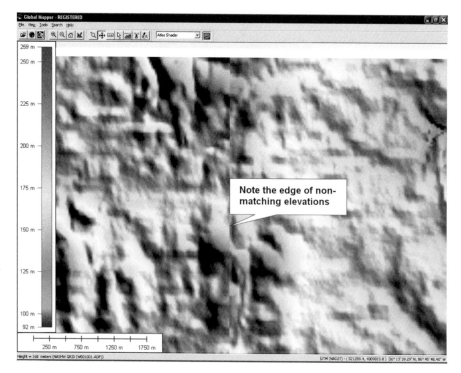

Fig. 4-4. Non-matching elevations along a boundary.

# DEM FORMATS

DEMs are usually delivered in either native DEM format or SDTS raster format. Native DEM format files consist of a single file usually with an extension of *.dem*. The file is in ASCII format and can be viewed with a text editor. In this format, the DEM consists of three types of records, denoted A, B, and C.

Record type A contains information about the entire file, including the file name, the southeast corner of the map, the DEM level (1, 2, or 3), the planimetric reference system, the zone if the coordinate system is UTM, projection parameters, the maximum and minimum elevation, the number of rows and columns, the horizontal and vertical datums, and the data source date. These, and other types of information, are all in one very long record (915 columns). Record type B contains the elevations and makes up the bulk of the file. Record type C contains, among other items, information on the RMSE and the number of sample points on which it is based.

> **NOTE:** *The details on how to read these records are in the appendices of* Standards for Digital Elevation Models, Part 2: Specifications, *available at* http://rockyweb.cr.usgs.gov/nmpstds/demstds.html.

Table 4-2 outlines some of the key items extracted from the DEM of Mount Rainier West.

**Table 4-2: Selected Information in Record Type A of a Native Format DEM**

| Item | Data |
| --- | --- |
| Name | MOUNT RAINIER WEST, WA |
| Process Code | 5 = DLG hypsography Linetrace |

| Item | Data |
|---|---|
| DEM Level | 2 |
| Sampling | 1 = Regularly Spaced |
| Planimetric Reference System | UTM Zone 10 |
| Ground Units | Meters |
| Elevation Units | Meters |
| Minimum Elevation | 731.6 meters |
| Maximum Elevation | 4393 meters |
| Data Source Date | 1998 |

Slide 10

In native format DEMs, information pertaining to the RMSE is reported as the last two digits in the DEM. For the Mount Rainier West 10-meter DEM the last two numbers listed are 2 and 30. These indicate that the RMSE is 2 meters based on a sample of 30 points (see slide 10).

The other major format for DEMs is SDTS raster. As with SDTS vector (discussed in Chapter 3), the SDTS raster format consists of a set of modules used to transfer information about the DEM. Software that reads and translates these files reports much, although not all, of the information outlined in Table 4-2.

# OBTAINING DEMS

DEMs can be obtained from many of the same sources listed for DRGs and DOQs. In particular, map libraries and state data centers are always good places to start. Additional resources are discussed in the following sections.

## USGS Business Partners' Free Download Sites

The USGS has contracted with at least three vendors: GeoComm International, MapMart, and Advanced Topographic Development and Images (see slides 11 through 13). These vendors have data available for free download, but the free links are sometimes slow. They may also have premium services that have faster data connections, but you pay per download or per CD you order. However, state and custom bundles are available, which may make using these sites worthwhile.

Slides 11–13

## USGS Sites

As with DLGs, DEMs are available through Earth Explorer (*http://earthexplorer.usgs.gov*). SDTS format DEMs are available at no charge through the three business partners listed in the previous section. The price for ordering through Earth Explorer is dependent on the format one chooses. For all formats and sizes, the price is $1.00 per file, plus a base charge ranging from $30 to $60 (see slides 14 and 15).

Slides 14–15

# WORKING WITH DEMS

Both Global Mapper and the ESRI products (ArcToolbox and ArcView 3.2) have modules for importing DEMs into a workable format. As you might expect, Global Mapper has several export options, whereas the ESRI tools create ESRI format files. In addition, there are some public domain tools available for creating point files from DEMs.

Global Mapper imports DEMs just like any other spatial data format it can handle. You do not have to un-*gzip* or un-*tar* any files. The program will recognize DEMs in either native or SDTS format. The program will report selected metadata about the DEMs loaded. You can

Slides 16–22

export the DEM to one of many raster formats. If there is more than one DEM loaded, the program has an option for filling gaps between adjacent DEMS. It will also smooth inconsistencies where the DEMs overlap but have different elevation estimates (see slides 16 through 22).

ArcToolbox has two modules for importing DEMs, depending on their format (see slides 23 through 26). For DEMs in native format, use DEM to Grid. For SDTS DEMs, use SDTS Raster to Grid (see figure 4-5). Both of these modules, along with the SDTS Import wizard, will create an ESRI grid, complete with selected metadata. As with vector SDTS files, the raster SDTS files must be un-*gzip*ped and un-*tar*red before you can use the translation tools. Chapter 3's warning concerning WinZip's un-*tar* options applies here, too.

Slides 23–26

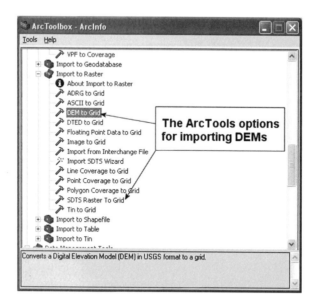

Fig. 4-5.
ArcToolbox
options for DEMs.

STDS2DEM and DEM2XYZ are programs that do not convert DEMs into GIS-ready formats. Rather, they extract out the X-Y-Z information and write it to a text file. These programs are public domain and can be downloaded from the GeoComm DEM web site (*http://data. geocomm.com/dem/*). They were written by the late Sol Katz, with updates by Greg Townsend of the University of Arizona. SDTS2DEM converts SDTS files to native format DEMs. DEM2XYZ converts DEMs to a text file consisting of X-Y-Z values (see slides 27 and 28). Figure 4-6 shows part of that file imported into ArcView 3.2. Such a file is instructional for understanding the structure and content of a DEM. Both programs are run from the command line, and thus you have to open a command window (DOS prompt window) to use them.

Slides 27–28

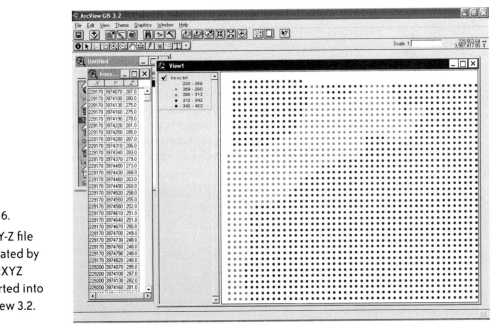

Fig. 4-6.

An X-Y-Z file generated by DEM2XYZ imported into ArcView 3.2.

# NATIONAL ELEVATION DATA

Tools for processing and making effective use of DEMs have improved over the years, and applications of DEMs have expanded into areas such as hydrologic analysis, line-of-site studies, views sheds, and cut-and-fill estimation.

However, users who want to use DEMs that cross zones or that have problems such as striping have had to spend a good deal of time on "data cleanup" and adjusting for different UTM zones. To better serve the user community, in the early 1990s the USGS began assembling a seamless elevation data set for the entire United States.

A single data set covering the United States, called the National Elevation Dataset (NED; see Gesh, D., M. Oimoen, S. Greenlee, C. Nelson, M. Steuck, and D. Tyler, *The National Elevation Dataset PE & RS*, vol. 68, pp. 5–32, 2002), was assembled from 7.5-minute DEMs available in 1999. This was a massive project that required edge matching and reprojecting nearly 57,000 7.5-minute quads. In the process of building the NED, systematic errors were addressed so that drainage systems based on the NED appeared more realistic. Figure 4-7 illustrates drainage networks derived from DEMs before and after various errors have been removed. The drainage pattern on the right is much more like those on the earth's surface.

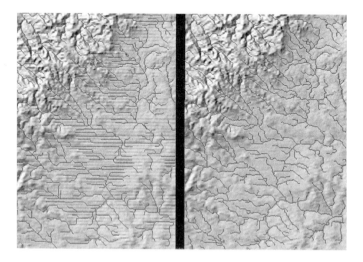

Fig. 4-7. Effect of error removal on drainage patterns (source: *http://gisdata.usgs.net/NED/About.asp*).

*TIP: A good discussion of the sources of errors in older DEMs and the main method for removing them can be found in An Effective Filter for Removal of Production Artifacts in USGS 7.5-minute DEMS (by Michael Oimoen), available at* http://gisdata.usgs.net/NED/paper.asp.

When building the NED, decisions had to be made concerning the projection to be used (Albers) and its horizontal datum (North American Datum of 1983). These differ from those used on most DEMs (UTM projection and North American Datum of 1927). Once the data set was built, the horizontal coordinate system was converted to decimal degrees. NED data come in 1-arc-second (approximately 30-meter) blocks, and for selected areas 1/3-arc-second (approximately 10-meter blocks). More 10-meter-resolution data are being added to the Seamless Download Data Site (*seamless.usgs.gov*) on a regular basis. The NED data are also available on CD or DVD.

## Obtaining NED Data

Slides 29–32

The NED data is free. You can download it from the Seamless Data Distribution viewer. The seamless viewer is part of the National Map program, which aims at providing users an easy way of obtaining data for the boundaries they specify (see slides 29 through 32).

The viewer uses web-based GIS to present an interactive map for downloading data sets. All elevation data on the Seamless Data Distribution System uses ESRI Grid format as the default distribution format. Other formats are BIL, BIP, TIFF, and GridFloat. BIL and BIP are binary files recognized by many GIS programs. Like the native format DEMs, they contain ASCII descriptor information. GridFloat is a non-proprietary floating point binary file with ASCII descriptor information. TIFF is the binary grid format discussed in Chapter 2.

## Working with NEDs

If you are using ESRI products, there is no need to import ESRI grid-format files. However, you will need to unzip the files that are downloaded from the seamless web site. Because at least two folders will be created, including an *INFO* folder, be sure to check *Use Folder names* when unzipping the data (so that the data does not end up in one folder). If you wish to work with another GIS system, Global Mapper (which supports several export formats) can read other NED-supported distribution formats.

You must take care when deriving surface characteristics such as hillshading and slope. The default of many programs is to assume that the X, Y, and Z coordinates are all in the same units. This is not the case in NED data, or in the SRTM data discussed in material to follow. If you proceed to calculate these surfaces without adjusting for the difference in horizontal and vertical units, the resulting surfaces will be incorrect. There are two options you can use to overcome this problem. The first is to reproject the NED data so that the X and Y values are in meters.

The second approach is to use a vertical exaggeration factor, or Z factor, parameter. The USGS used 0.00003 as the Z factor parameter when creating a shaded relief map of the United States. However, it may be more accurate to calculate a factor, particularly for slope calculations. Recall that for 36° (the approximate latitude of Knox County), 1 degree of longitude is approximately 90053.52 meters. Put another way, an appropriate Z factor for 36° North longitude would be 1/90053.52 (or 0.000011). You may want to calculate the appropriate Z factor for the median longitude of your study area. If your study area has a large north-south extent and you wish to calculate slopes, reprojecting the NED data is probably your best option. An example of entering such a conversion factor in ArcMap is shown in slide 33.

Slide 33

# SHUTTLE RADAR TOPOGRAPHY MISSION DATA

On February 11, 2000, the Shuttle Radar Topography Mission was launched. Lasting just over 11 days, the mission used radar interferometry to map the topography of North and South America. Radar signals were bounced off the earth and collected on two antennas, one in the shuttle's cargo bay and the other on a 60-meter mast. By comparing the reflectance patterns and knowing the distance between the antennas, the elevation of the earth's surface was calculated. A discussion of this process can be found at *http://srtm.usgs.gov*.

The SRTM is a joint project of the National Aeronautics and Space Administration and the National Geospatial-Intelligence Agency. The German Aerospace Center and the Italian Space Agency were also sponsors of the mission. The goal is to produce a digital representation of topography for all land areas between 56° South and 60° North latitude. Data points are to be located every 1 arc second, or approximately 30 meters. The data is distributed by the USGS. For the conterminous United States, the data are available at 30- and 90-meter resolution. For the rest of North America and South America, they are available at 90-meter resolution.

Radar has several advantages over photography. It operates in both daylight and darkness. In addition, it can "see" through cloud cover. Since the measurements are taken from space, overflight permission of other countries is not required. Currently, SRTM data is available at 30-meter resolution for most of the United States at 60° North or lower latitude. 90-meter resolution SRTM data is available for the rest of North and South America that lies between the northern and southern boundaries of the data collection extent.

You might think that 30-meter NED data will match up nicely with 30-meter SRTM data. This is not necessarily the case. Each data set uses a different datum, and thus there will be differences in the estimates of elevation at the same location. The PowerPoint slides for this chapter illustrate the differences between 90- and 30-meter SRTM and 30- and 10-meter NED data (see slides 34 and 35).

Slides 34–35

## Obtaining and Using SRTM Data

The SRTM data can be downloaded free of charge from the USGS Seamless Data Distribution System web site. The process is just like downloading NED data, except that you must make sure the SRTM entries are checked in the download layers section. There are two viewers that can be used when working with SRTM. For areas outside the United States (currently limited to the Americas), use the international viewer. For the United States, you can use either the national or the international viewer. The data can be downloaded in the same formats as NED (ArcGrid, BIL, TIFF, and GridFloat). The same program considerations that apply to NED data apply here. Warnings about hillshade and slope calculations that were made for NED data also apply to the SRTM. That is, you must adjust for the fact that the horizontal units are in decimal degrees and the vertical units in meters.

## Combining Data Sets

Slides 36–37

Chapters 2 through 4 have covered three types of data—raster images (DRGs and DOQQs), vector layers (DLGs), and DEMs—that for certain scales can be easily combined. An interesting exercise is to combine these data sets. This can yield much richer visualizations than can be found by viewing each data set on its own (see slides 36 and 37). For example, figure 4-8 shows a hydrologic DLG layer draped over a DEM. The combination clearly illustrates the relationship between elevation and drainage.

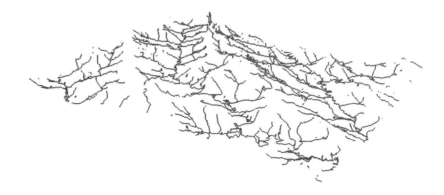

Fig. 4-8. Hydrologic DLG draped on a DEM.

# WEB SITES AND DOCUMENTATION

Slides 38–40

The USGS has many web pages and documents that pertain to DEM and NED data sets. These include fact sheet pages, product description pages, user guides, and National Mapping Standards documents. The URLs for these sites are listed in slides 38 through 40 of the PowerPoint file for this chapter. Figure 4-9 shows the fact sheet for DEMs, and figure 4-10 shows the NED data product page.

Because the NED is derived from DEMs, statements of accuracy about DEMs also apply to NEDs (see *http://gisdata.usgs.net/NED/AccuracyQ1.asp.*) The accuracy of these data sets is dependent on the scale and quality of the source data. As work on the NED continues, the NED will contain information based on 7.5-minute and 15-minute DEM products. You can find the source data used in the NED at the Data Source Index interactive web map (*http://gisdata.usgs.net/website/USGS_GN_NED_DSI*).

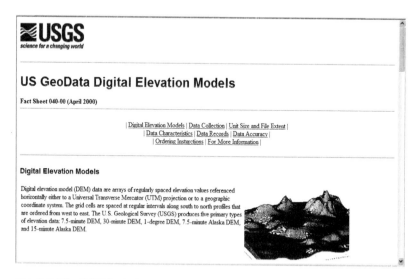

Fig. 4-9. The DEM fact sheet.

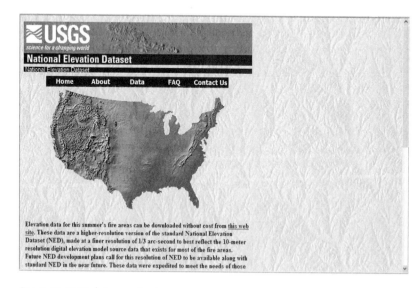

Fig. 4-10. NED data product page.

Slide 41

The USGS-based SRTM web sites are listed in slide 41 of the PowerPoint file for this chapter. There is an SRTM home page, a mission description page, coverage map, and fact sheet. In addition to these USGS-maintained pages, NASA's Jet Propulsion Laboratory (JPL) has a web site of images based on SRTM relief and Landsat images (see figure 4-11). When combined, the elevation and satellite image data can yield impressive views of the earth. Figure 4-12 shows the Los Angeles area from space. It is even more impressive in color (see slide 42). A gallery of images from around the world can be found at *http://www2.jpl.nasa.gov/srtm/dataprod.htm#Gallery*. The Landsat data used to generate the JPL's images serves as the basis for the National Land Cover Database. Land cover is explored in Chapter 5.

Slide 42

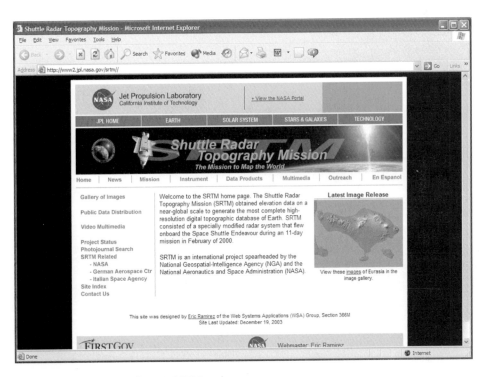

Fig. 4-11. The JPL SRTM web page.

Fig. 4-12. The Los Angeles basin generated by combining SRTM elevations with Landsat imagery (image courtesy of NASA/JPL/CalTech).

# CHAPTER 5

# LAND COVER DATA

## INTRODUCTION

THIS CHAPTER EXPLORES LAND COVER DATA. There are some parallels with topics covered in previous chapters. In Chapter 4, you were exposed to two elevation data sets, one of which was developed primarily from NAPP source materials or topographic map data (DEMs) and the other (SRTM) from radiometry taken from space. Raster data sources (in particular, DRGs) were covered in Chapter 2, and vector representations of topographic maps (DLGs) were the focus of Chapter 3. Land cover data show how land data is used and what is on the landscape, as revealed from aerial photographs and satellite imagery. Land cover data can be used to study tree health in areas of wildfires, areas of drought, seasonal and climatic changes, urban growth, and other phenomena.

Two national land data sets exist: Land Use Land Cover (LULC) data and the National Land Cover Database (NLCD). The LULC was developed from aerial photography, land-use maps, and surveys, whereas the NLCD was developed from data collected from space. The LULC is a vector representation of land use and land cover, whereas NLCD (based primarily on Landsat 7 imagery) is a raster data set. Both represent land cover patterns, but the NLCD is more recent and a new version of the NLCD should be released in 2004. Table 5-1 presents an overview of the two land cover data sources.

**Table 5-1: LULC and NLCD Overview**

| Item | Land Use and Land Cover | National Land Cover Dataset |
|---|---|---|
| Description | Land cover classification. | Land cover classification. |
| Format | Vector format in GIRAS (EPA distributes ArcInfo GRID format files). | Raster format. |
| Scale and scope | Most of the United States at 1:100000 and 1:2500000 (some missing areas in Alaska). Minimal polygons of manmade areas are 10 acres; 40 acres for natural features. Width limits apply, too. Coverage is not uniform across the country. | United States at 30-meter resolution. |
| Projection and datum | UTM in zone of corresponding topographic map, NAD83. | Decimal degrees in NAD83. |
| Currency | Last updated in 1980s. | 1992; a new version due out in 2004. |
| Source | Aerial photos, topographic maps, land-use maps, and surveys. | Landsat 7. |
| Software considerations | Workstation ArcInfo and Global Mapper. | Similar to other raster data sets. |
| Other considerations | Land cover based on Anderson classification scheme. | Land cover based on modified Anderson classification scheme. |

# LAND USE AND LAND COVER DATA

Land Use Land Cover data sets provide vector representations of land cover for most of the United States. During the 1970s and 1980s, aerial photography was interpreted to delineate areas of similar land use. The main sources of information for the LULC data are National High Altitude Aerial Photography (NHAP) photographs, land-use maps, and field surveys. These photos were studied to delineate areas corresponding to specific land-use classifications.

The land-use classifications were based on a hierarchical system developed by Anderson et al. in 1976. (You can download the original paper at *http://landcover.usgs.gov/pdf/anderson.pdf*.). Their system was based on a series of coded values. Each value was based on a set of digits. The more digits, the more specific the land-use category. For example, a value of 1 indicated an urban or built-up area. Types of built-up areas included 11 for residential, 12 for commercial and services, 13 for industrial, and so on. These are listed in Table 5-2.

**Table 5-2: Anderson Classification System at the Two-digit Level**

| Code | Category |
| --- | --- |
| 11 | Residential |
| 12 | Commercial and services |
| 13 | Industrial |
| 14 | Transportation, communications, and utilities |
| 15 | Industrial and commercial complexes |
| 16 | Mixed urban or built-up land |
| 17 | Other urban or built-up land |

| Code | Category |
| --- | --- |
| 21 | Cropland and pasture |
| 22 | Orchards, groves, vineyards, nurseries, and ornamental horticulture |
| 23 | Confined feeding operations |
| 24 | Other agricultural land |
| 31 | Herbaceous rangeland |
| 32 | Shrub and brush rangeland |
| 33 | Mixed rangeland |
| 41 | Deciduous forest land |
| 42 | Evergreen forest land |
| 43 | Mixed forest land |
| 51 | Streams and canals |
| 52 | Lakes |
| 53 | Reservoirs |
| 54 | Bays and estuaries |
| 61 | Forested wetland |
| 62 | Nonforested wetland |
| 71 | Dry salt flats |
| 72 | Beaches |
| 73 | Sandy areas other than beaches |
| 74 | Bare exposed rock |
| 75 | Strip mines, quarries, and gravel pits |
| 76 | Transitional areas |

| Code | Category |
|------|----------|
| 77 | Mixed barren land |
| 81 | Shrub and brush tundra |
| 82 | Herbaceous tundra |
| 83 | Bare ground tundra |
| 84 | Wet tundra |
| 85 | Mixed tundra |
| 91 | Perennial snowfields |
| 92 | Glaciers |

For an area to be classified with a given land use it has to meet certain minimal size limits. For certain categories (11 through 17, 23, 24, 51 through 54, and 75 and 76) the area must be at least 4 hectares in size. All other categories are based on a minimal size of 16 hectares. In addition, urban or built-up areas (11 through 17) and water categories (52 through 54) have to have a minimal width of 200 meters. Lower widths are acceptable for limited-access highways (category 14) and double line rivers (51). Those must have a minimum width of 92 meters.

Manual interpretation of aerial photography over two decades has resulted in the development of a vector-based LULC database. Most of the country is covered at the 1:250000 scale. However, gaps exist; for example, only the area around Valdez is available for Alaska. At the 1:100000 scale, the coverage is even more limited. You can go to *http://edc.usgs.gov/geodata/* and click on LULC to retrieve lists of LULC data at each scale (see slides 1 through 4). All LULC data is in the UTM projection and should closely match with corresponding topographic maps. The majority of LULC data corresponds with 1:250000 USGS topographic maps.

Slides 1–4

Slides 5–9

As with DLGs, LULC vector files are classified into groups. These are Land Use, Political, Census, Hydro, and Federal. Sample maps of these are in the PowerPoint file for this chapter (see slides 5 through 9). The Land Use theme consists of polygons with attribute codes listed in Table 5-2. The Political theme consists of outlines of states, counties, and independent cities. Depending on the date of the source material, the Census theme consists of standard metropolitan statistical areas (SMSAs) or metropolitan statistical areas or primary metropolitan statistical areas (MSA/PMSA), along with Census county subdivisions. As will be discussed in Chapter 6, TIGER files are a better source for this information and are more up-to-date than the LULC maps. (However, the TIGER files are in decimal degrees, not UTM coordinates.)

Polygons are defined by eight-digit codes in the Hydro theme units. The codes indicate the region (digits 1 and 2), subregion (digits 3 and 4), accounting unit (digits 5 and 6), and cataloging unit (digits 7 and 8) for each hydrologic unit. Figure 5-1 illustrates how the codes are to be interpreted. The selected hydrologic unit has a code of 06010202. (Note that ArcInfo drops the leading 0 when it imports the GIRAS file.) The 06 indicates that this is in an area in Tennessee.

The next two digits (01) specify the subregion as the Upper Tennessee River Basin. The 02 that follows indicates that it is the Upper Tennessee excluding the French Broad and Holston River basins (the accounting unit). Finally, the last 02 indicates that this feature is the Upper Little Tennessee River. A complete discussion of the codes, along with a map of the regions, can be found at *http://water.usgs.gov/pubs/circ/circ878-A/html/pdf. html.*

The more recent NHD data, discussed in Chapter 8, is a better source of information on watersheds. The NHD

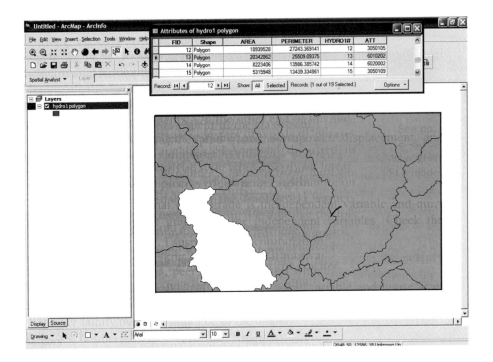

Fig. 5-1. Hydrologic LULC layer.

also uses the eight-digit hydrologic unit codes. The Federal theme contains areas of over 16 hectares that are owned or managed by the federal government (see slide 10). Ownership of federal lands is indicated by a two-digit code. Table 5-3 outlines federal land ownership codes and the agencies they represent.

Slide 10

## Table 5-3: Federal Land Ownership Codes

| Code | Agency |
|------|--------|
|      | **Department of Agriculture** |
| 11   | Agricultural Research Service |
| 12   | Forest Service (National Forest) |
| 13   | Forest Service (National Grassland) |

| Code | Agency |
|------|--------|
|      | **Commerce Department** |
| 21   | National Oceanic and Atmospheric Administration |
|      | **Defense Department** |
| 31   | Air Force |
| 32   | Army |
| 33   | Army (Corps of Engineers - Civil Works) |
| 34   | Navy |
|      | **Department of the Interior** |
| 41   | Bonneville Power Administration |
| 42   | Bureau of Indian Affairs (does not include Indian lands held in trust) |
| 43   | Bureau of Land Management |
| 44   | Bureau of Mines |
| 45   | Bureau of Reclamation |
| 46   | Fish and Wildlife Service (National Wildlife Refuge) |
| 47   | National Park Service (National Monument, Seashore, and Recreation Area) |
| 48   | National Park Service (National Park) |
|      | **Justice Department** |
| 51   | Bureau of Prisons |
|      | **Department of State** |
| 61   | International Boundary and Water Commission, United States and Mexico |
|      | **Department of Transportation** |
| 71   | Federal Aviation Administration |

| Code | Agency |
|------|--------|
| 72 | Federal Railroad Administration |
| 73 | U.S. Coast Guard |
| | **Other Agencies** |
| 81 | Energy Research and Development Administration |
| 82 | General Services Administration |
| 83 | National Aeronautics and Space Administration |
| 84 | Tennessee Valley Authority |
| 85 | Veteran's Administration |

# OBTAINING AND WORKING WITH LULC DATA

You can freely download LULC data sets from the USGS geodata site (*http://edc.usgs.gov/geodata*). They are also available from the Environmental Protection Agency, but only at the 1:2500000 scale. The EPA distributes the LULC data as ArcInfo interchange (E00) files. The USGS distributes these files in either GIRAS (Geographic Information Retrieval and Analysis System) or CTG (Composite Theme Grid) formats. However, only the GIRAS format was available from the FTP site at the time of this writing.

Working with GIRAS files from USGS requires some preprocessing. GIRAS-formatted data comes as zipped files. The records within those files do not contain carriage returns. GIRAS translators expect records of 80 characters in length. However, when unzipped the resulting file has only one record, with its length equal to the number of character spaces in the file. The data must be "chopped" into 80-character-length records. This is done by one of two methods. On UNIX systems, there is a simple command for accomplishing this.

On PC systems, you can use a freeware program called CHOP.EXE. The program is available at *http:// edc.usgs.gov/geodata/public.html*. This must be done before the GIRAS data can be processed by the manipulation programs discussed in material to follow (Global Mapper or Workstation ArcInfo). An illustration of how the CHOP program works can be found in the PowerPoint slides for this chapter (see slides 11 through 13). Once the file downloaded from the USGS site is properly parsed into records of 80 characters in length, you can then process it with either Global Mapper or with Workstation ArcInfo.

Slides 11–13

Global Mapper does not automatically recognize GIRAS-format LULC files. You must explicitly indicate the type of file during the Global Mapper Open File process (see figure 5-2). After importing the LULC file, Global Mapper will then assign a rendering based on the LUCODE values. Since this is a vector data set, you can export to any vector format Global Mapper supports. For example, if you export to a shapefile, the polygon theme will have a character field describing the land-use type for each polygon.

Fig. 5-2. Importing GIRAS files in Global Mapper.

ArcToolbox does not support GIRAS imports. However, Workstation ArcInfo does. The Workstation ArcInfo command GIRASARC converts the GIRAS file into an ESRI coverage (see figure 5-3). After importing the LULC file, you will have to create polygon topology via either the CLEAN command or the BUILD with POLY option. The resulting coverage will have an attribute field containing the LUCODE values. This can be joined to the Anderson.DBF file (found on the companion CD-ROM) to generate a thematic map of land use and land cover (see figure 5-4). Details on joining the coverage PAT file with the Anderson attributes can be found in the PowerPoint file for this chapter (see slides 14 through 18).

Slides 14–18

In addition to the GIRAS data described previously, LULC data is available from the EPA as ESRI export files, which are also called e00 files (see slide 19). Once unzipped, ESRI products such as ArcView 3 and ArcToolbox can import these files into coverage formats (see figure 5-5). EPA data is available only at the 1:250000 scale.

Slide 19

```
Arc                                                              - □ ×
Arc: girasarc land_use.grs wknox

** Converting land_use.grs into wknox **
Number of Arcs in Map.........      15964
Number of Points in Map.......     183432
Number of Polygons in Map.....       6562
Number of Text Records........          0
Point Tolerance = 2, Arc Length Tolerance = 7
Map Type = 1  Land Use and Land Cover
Map Projection Code = 1, Map Scale Denominator = 393700
Source Map Date - 1976, File Created - 82159
Map Title: KNOXVILLE, TN SC GA 1:250000 QUAD LAND USE MAP

Arc: describe wknox
            Description of SINGLE precision coverage wknox

                        FEATURE CLASSES

                              Number of   Attribute    Spatial
Feature Class       Subclass  Features    data (bytes) Index?    Topology?
--------------      --------  ---------   ------------ -------   ---------
ARCS                            15964
POINTS                           6562          16

                        SECONDARY FEATURES
```

Fig. 5-3. Using GIRASARC.

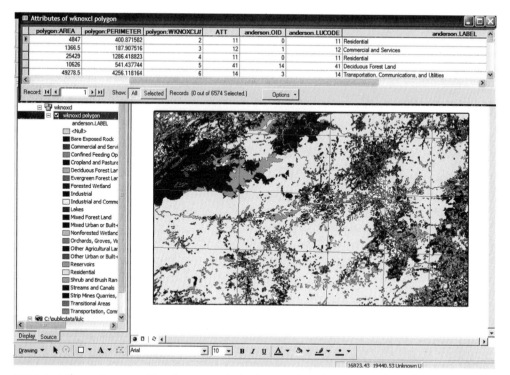

Fig. 5-4. A thematic map of land cover with Anderson classifications

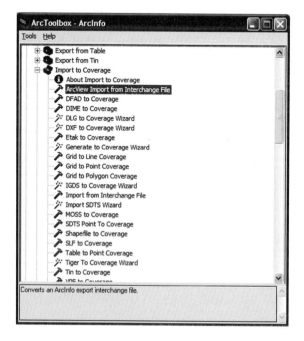

Fig. 5-5. Import
option of e00
vector files.

# NATIONAL LAND COVER DATA

The LULC data was compiled from aerial photography during the 1970s and 1980s. It is, therefore, considered a historic record of land use rather than a current one. A seamless NLCD set based on Landsat Thematic Mapper imagery from the early to mid 1990s, along with supplementary data sources, was completed in September of 2000. Often referred to as NLCD 92, this raster-based land cover data set was developed from Landsat images taken from 1991 to 1993. Like the National Elevation Dataset discussed in the previous chapter, the NLCD consists of raster data in decimal degrees based on the North American Datum of 1983.

The NCLD is supported by a consortium of federal agencies, including the USGS, the EPA, the U.S. Forest Service, and the National Oceanic and Atmospheric Administration. Together this group of agencies is referred to as the Multi-Resolution Land Characterization Consortium (see slide 20). The MRLC is supporting a current project to update this map with 2001 Landsat ETM+ imagery. The expected completion date is sometime in 2004.

Slide 20

The NLCD land cover categories are based on a modified Anderson scheme. A two-digit code indicates each land cover type. Table 5-4 lists each code and its meaning.

**NOTE:** *A discussion of the similarities and differences between the Anderson system and the NLCD system can be found at* http://landcover.usgs.gov/classes. html.

**Table 5-4: NLCD Land Cover Codes**

| LULC Code | Meaning |
| --- | --- |
| 11 | Open water |
| 12 | Perennial ice and snow |
| 21 | Low-intensity residential |
| 22 | High-intensity residential |
| 23 | Commercial/industrial/transportation |
| 31 | Bare rock/sand/clay |
| 32 | Quarries/strip mines/gravel pits |
| 33 | Transitional |
| 41 | Deciduous forest |
| 42 | Evergreen forest |
| 43 | Mixed forest |
| 51 | Shrubland |
| 61 | Orchards/vineyards/other |
| 71 | Grasslands/herbaceous |
| 81 | Pasture/hay |
| 82 | Row crops |
| 83 | Small grains |
| 84 | Fallow |
| 85 | Urban/recreational grasses |
| 91 | Woody wetlands |
| 92 | Emergent herbaceous wetlands |

## Land Cover Classification Methods

The NLCD uses at least two Landsat images: one with deciduous vegetation leaves off and one with leaves on. Four TM bands (3, 4, 5, and 7) are combined and subjected to a statistical process known as unsupervised cluster analysis. Essentially, this process groups together those pixels that have similar patterns on the four bands. The result is an image consisting of 100 classes. These are then assigned to one of the 21 categories listed in Table 5-4. The classification is then tested against NAPP information and adjustments are made using supplementary data.

Any statistical grouping of pixels into like groups will yield some misclassifications, or groupings. For example, it may be difficult in urban areas to determine if an area is high-density residential or commercial (classes 22 or 23 in Table 5-4). By using population density data from the U.S. Census, the land cover can be more accurately determined. The previously cited urban example is probably not subject to seasonality constraints. That is, a built-up area is built up whether leaves are on or off. The same cannot be said of other classifications.

Forests present a case where seasonality is crucial. Studying an area in the leaves-on image may indicate that it is forested, but it is the leaves-off image that differentiates between evergreen and deciduous forests. Thus, the analysis of these land cover images is time sensitive. That is, the differences between seasons are crucial. For other land cover categories, timing *within* a season is critical.

Consider the Pasture/Hay and Row Crops classifications. During the leaves-on season both of these areas have a high degree of greenness. During the spring, however, there is a short window of opportunity where pasture/ hay areas "green up" before row crops areas. If the Landsat imagery takes place during this window, differentiating between these two classes is manageable. However, if the image is either too early or too late, the probability of confusing these areas is high. Considering elevation and slope from digital elevation data can lower this probability.

To address these and other uncertainties resulting from the statistical clustering of TM bands, the NLCD uses ancillary data from the Census Bureau, 3-arc-second (30-meter) DTM data (including derived surfaces such as slope and aspect), LULC data, and the National Wetlands Inventory. Wetlands, in particular, are difficult to classify with Landsat images, and thus the NWI is necessary for accurately classifying the type of wetland.

> **NOTE:** *For a more detailed discussion of the classification issues and procedures, visit the Mapping Procedures web site at* http://landcover.usgs.gov/ mapping_proc.html.

## Accuracy of the NLCD

Slide 21

Once the pixels derived from 30-meter Landsat data are classified, their accuracy is tested against information from the NAPP (see slide 21). However, the NAPP is much more detailed, being at 1-meter resolution (see figure 5-6). This difference in resolution presents a challenge when determining the accuracy of the NLCD.

Because of the difference in resolution, there are several ways to assess agreement of land cover classifications between the NLCD and the NAPP data. One method would be to determine if the pixel in the NAPP that was at the center of the NLCD grid cell matched the classification of the NLCD. This is referred to as a single pixel match. The second method would be to see if the most common classification of pixels in the NAPP within a predefined neighborhood of the NLCD matches the NLCD classification. This is referred to as the mode match.

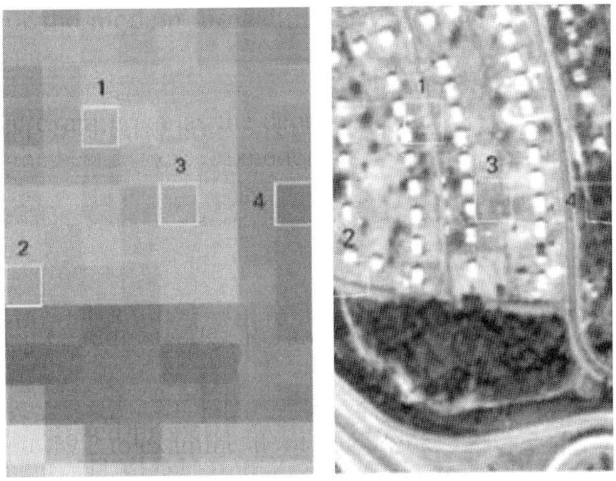

Fig. 5-6. Comparing NCLD and NAPP information (source: *http:// landcover.usgs.gov/ accuracy/figure2.html*).

The mode match is probably the better way to measure accuracy because it explicitly recognizes the need to generalize as the size of the spatial unit increases. A comparison of the measures of matching is presented in the PowerPoint file (see slides 22 and 23) for this chapter. The results from the matching tests indicated that closely related land cover classes were most likely the sources of confusion. Examples include confusing open water with emergent wetland, deciduous or evergreen forests with mixed forests, and high-intensity residential with high-intensity commercial properties.

Slides 22–23

> **NOTE:** *For a complete list of identified confusions by region of the country, see* http://landcover.usgs.gov/accuracy/table6. html.

## Obtaining NLCD Data

NLCD 92 data is available by state or as part of a national seamless data set (*seamless.usgs.gov*) available through the National Map (see slides 24 through 27). State data consists of the national data that were clipped by state

Slides 24–27

boundaries as defined by the 1:100000 DLG boundary files. Because there may be some discrepancies with actual boundaries and the 1:100000 DLGs, a 10-pixel buffer (300 meters) was used around each state.

> **NOTE:** *State data is available at* http://edc.usgs.gov/ geodata. *Click on the NLCD link and follow the Products link. The NCLD data for each state is available in GeoTIFF format.*

NLCD for any part of the conterminous United States and Hawaii can be accessed through the seamless data download site. As discussed previously, that site supports several formats. For the NLCD, the available formats are GeoTIFF, BIL, and ArcGrid. The steps for downloading the NLCD are similar to those for elevation data at that site (e.g., National Elevation Data and Shuttle Radar Topography Mission). These steps were discussed in the previous chapter.

> **NOTE:** *Statistics on land cover are available at* http:// landcover.usgs.gov/nlcd.html. *From this page, you can find statistics on the areas of various land cover classifications by state or region. Other useful web sites exist in the PowerPoint file for this chapter.*

## Using NLCD with NED Data

NLCD and NED data are often combined to produce 3D renderings of land cover maps or to add hillshading to the land cover map. The companion CD-ROM contains a file named *NLCD.dbf,* which consists of a lookup table of land cover classifications. The content of the file is shown in figure 5-7. By joining an NLCD raster data set to the lookup table, you can assign meanings to each value of the NLCD raster data set. Slide 28 of the PowerPoint file for this chapter illustrates such a use. When using NLCD with elevation data, it is easiest to use it with the NED

Slide 28

data, in that they both have the same coordinate system, datum, and spheroid.

| OID | LUCODE | MEANING |
|---|---|---|
| 0 | 11 | Open Water |
| 1 | 12 | Perennial Ice_Snow |
| 2 | 21 | Low Intensity Residential |
| 3 | 22 | High Intesity Residential |
| 4 | 23 | Commercial_Industrial_Transporta |
| 5 | 31 | Bare Rock_Sand_Clay |
| 6 | 32 | Quarries_Strip Mines_Gravel Pits |
| 7 | 33 | Transitional |
| 8 | 41 | Deciduous Forest |
| 9 | 42 | Evergreen Forest |
| 10 | 43 | Mixed Forest |
| 11 | 51 | Shrubland |
| 12 | 61 | Orchards_Vineyards_Other |
| 13 | 71 | Grasslands_Herbaceous |
| 14 | 81 | Pasture_Hay |
| 15 | 82 | Row Crops |
| 16 | 83 | Small Grains |
| 17 | 84 | Fallow |
| 18 | 85 | Urban_Recreational Grasses |
| 19 | 91 | Woody Wetlands |
| 20 | 92 | Emergent Herbaceous Wetlands |

Fig. 5-7. Content of the *NLCD.dbf* lookup table.

## NLCD 2001

An updated version of the NLCD, NLCD 2001, is being created by the Multi-Resolution Land Characterization 2001. This will use Landsat 7 imagery along with ancillary data to create a second-generation NLCD data set. The new data set will cover all 50 states and Puerto Rico.

> **NOTE:** *The strategy to be used and the lessons learned from building the 1992 data set can be found in Homer et al.,* Development of a Circa 2000 Landcover Database for the United States, *available at* http://landcover.usgs.gov/pdf/asprs_final.pdf.

# WEB SITES AND DOCUMENTATION

Product pages exist for both the LULC and NLCD data sets. The LULC product page (*http://edc.usgs.gov/products/*

*landcover/lulc.html*) contains a brief product description, pricing information (the data is free), and links to USGS and EPA download sites (see figure 5-8). If you download the LULC information from the USGS geodata site (*http://edc.usgs.gov/geodata*), you will have to follow the steps for preparing GIRAS data (discussed previously). The EPA site organizes the LULC information into three areas: east of 86 degrees west longitude, between 86 and 110 degrees west longitude, and west of 110 degrees west longitude.

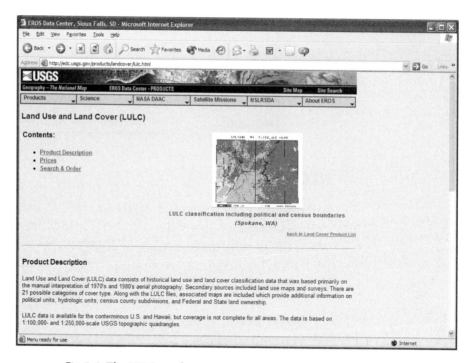

Fig. 5-8. The LULC product page.

The exported ArcInfo coverage formats (e00 files) are *gzipped*. Once unzipped, the E00 file can be imported into ArcGrid coverages using ESRI import tools (see figure 5-5). The files available from the EPA, which are at the 1:250000 scale, have a naming convention similar to that used for DRGs. Each file name begins with the lowercase letter *l*, followed by the first two letters of the 1:250000 quadrangle. The next two digits are the latitude, followed by three digits for the longitude of the southeast corner of the quad. For example, the file named *lkn35082.e00.gz* is the zipped E00 file for the Knox quadrangle with southeast corner at 35 degrees north and 82 degrees west. If you wish to learn more about the source data used for creating the LULC, visit the NHAP web sites discussed under "Web Sites and Documentation" at the end of Chapter 1.

There are several web pages associated with the NLCD. The Land Cover Characterization Program page (*http:// landcover.usgs.gov*) describes the LCCP, of which the NLCD is part. The program page lists various projects that require land cover characterization information (see figure 5-9). Examples include urban dynamics, land cover trends, biodiversity analysis, and gap analysis.

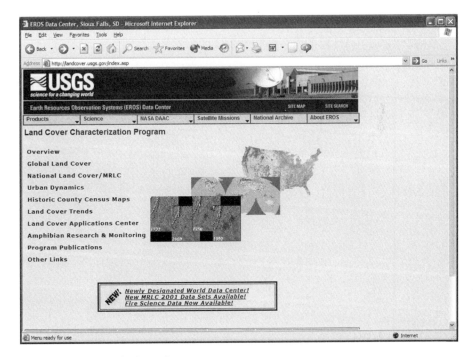

Fig. 5-9. The LCCP home page.

Slide 29

There are two NLCD project pages: one for the original NLCD and one for the 2001 NLCD. The URLs for these sites are listed on slide 29 of the PowerPoint file for this chapter. The 2001 page contains more detailed information on the NLCD 2001 database content and methodology. Links lead to publications that offer in-depth discussion of the NLCD process (see *http://landcover. usgs.gov/publications.asp*). The 2001 NLCD, which has a target completion date of 2004, divides the United States into mapping zones (see figure 5-10). The zones are determined using five criteria: economies of size, physiography, land cover distribution, spectral uniformity, and edge matching of Landsat images. A more detailed discussion of the partitioning of the United States into zones can be found in "Partitioning the Conterminous United States into Mapping Zones for Landsat TM Land Cover Mapping," by Homer and Galant, available at *http://landcover. usgs.gov/pdf/homer.pdf*.

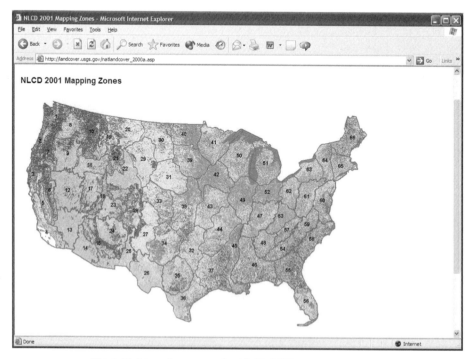

Fig. 5-10. Mapping zones for NLCD 2001.

CHAPTER 6

# TIGER DATA

## INTRODUCTION

TOPOGRAPHICALLY INTEGRATED GEOGRAPHIC ENCODING and Referencing (TIGER) files contain information about the boundaries of Census enumeration units, and on several types of point and line features. Whereas previous chapters dealt with data sets provided primarily by the United States Geological Survey (USGS), this chapter and Chapter 7 focus on data provided by the United States Census Bureau. Chapter 7 focuses on the attribute data collected by the Census Bureau, whereas the focus of this chapter is on the geographic files produced by the Census Bureau. Table 6-1 provides an overview of TIGER files.

**Table 6-1: TIGER Overview**

| Item | Description |
|------|-------------|
| Description | Files that define census-related polygons. These are needed for mapping census data. Information on various line features, particularly roads, is used in geocoding. |
| Format | TIGER files come in their own format. Information is spread across many files (the number depends on the version of TIGER). Each file represents a record type. |

| Item | Description |
|------|-------------|
| Projection/ coordinate system | TIGER data coordinates are in decimal degrees of latitude and longitude. Since 1995, TIGER files have been based on NAD83. From 1988 to 1994, the datum used was NAD27. |
| Original source | The original source for TIGER was a combination of the Census Bureau's Geographic Boundary Files/Dual Independent Map Encoding (GBF/DIME) source and 1:100000-scale USGS DLGs. However, many sources have now contributed to TIGER's development. Various local agencies supply the Census Bureau with information, and thus there are no longer just two sources for TIGER. |
| Software considerations | Translating TIGER requires using tools such as TGR2SHP or the ESRI TIGER translator. The degree of knowledge necessary about the relationship between the many record types varies according to the translator. I have tried to make TGR2SHP as intuitive as possible. |
| Other considerations | TIGER files change regularly and these changes include format changes. As a result, translation software should be updated with each TIGER release. |
| Software used in this chapter | TGR2SHP, ArcToolbox, TIGER translator. |

Slides 1–2

There are several key points of which you should be aware when embarking on a census-based GIS project (see slides 1 and 2). TIGER files do not contain the population and housing counts from the Census. That is, data collected in a particular decennial census are not part of the TIGER files available from the U.S. Census Bureau. TIGER files contain geographic information and attributes about census geography, not counts. In addition, raw TIGER files are not GIS-ready. In fact, a TIGER "file" is really a collection of files with each file containing a specific record type. Translation software must be used to extract the desired geography from those files and write it in the desired format.

TIGER is distributed by county. To assemble layers for areas that cross county lines, such as American Indian areas and metropolitan areas, you may need to combine information from several counties. Fortunately, modern extraction and GIS software makes this fairly easy. Finally, TIGER files are updated frequently. In just over a decade, for example, the Census Bureau has released the following versions of TIGER: TIGER 92, TIGER 94, TIGER 95, TIGER 97, Local Update of Census Address TIGER, TIGER 98, Census 2000 Dress Rehearsal TIGER, TIGER 99, Redistricting TIGER, TIGER 2000, TIGER 2000 Urbanized Areas, TIGER 2002, and 108th Congressional Districts TIGER. Each version of TIGER was closely related to previous versions, but each contained different enumeration units, area definitions, address ranges, updated streets, and so on.

Each version of TIGER used slightly different file structures, and the Census Bureau changed the datum, starting with TIGER 95, to NAD83. Prior to that version of TIGER, the datum used was NAD27, with regional datums for Hawaii and the Pacific. Starting with 2002, there were some significant additions to TIGER content and file structures, with more changes planned for the rest of this decade.

TIGER files grew out of the Geographic Boundary File/Dual Independent Mapping and Encoding (GBF/DIME) program developed for the 1970 Census and expanded and updated for the 1980 Census. These files existed for urban areas, but in order to expand the system to cover the entire nation, USGS DLGs were created to fill in rural areas. This combination of data sets, leading to the release of TIGER files in 1988, supported the collection of census data. It is fair to say that the GBF/DIME program did much to popularize the topological vector model in GIS. More to the point, TIGER was not developed to support extremely accurate cartography, hydrologic analysis, or other such applications. If you try to use TIGER for these purposes, as many users have, you will most likely be disappointed with your results.

TIGER should be used for its intended purpose: to support the collection and display of Census data. It can also support address matching, particularly in urban areas. Specific initiatives have begun to improve TIGER's positional accuracy and the quality of some of its layers. For example, efforts are underway to use aerial photography to improve the location of features (much like what is done for topographic maps), to develop more complete and accurate address ranges, to incorporate hydrologic information from the National Hydrography Dataset (see Chapter 8), and to improve topological information to account for overpasses and underpasses. Let's take a look at why such changes are needed.

# SPATIAL ACCURACY

Relative positions in TIGER are consistent. That is, a map of an area using only TIGER layers will look topologically correct. However, if you overlay certain line layers on a DLG, DOQQ, or 1:100000-scale DRG, there will be noticeable discrepancies between the orthophoto and the TIGER line features. The PowerPoint file for this chapter contains the roads and hydrology line features from TIGER that have been projected into the same coordinate system and datum as a 1:100000-scale DRG (see slides 3 through 5).

Slides 3–5

The resulting discrepancies are clear. The line segments in TIGER are less smooth than the DLGs. In addition, major discrepancies in location are common, and the roads that are contained in a DLG will more closely adhere to their true location. Compare the slide of TIGER roads and a DRG with the corresponding DLG road slide in slide 13 in Chapter1's PowerPoint file. The Census Bureau has embarked on an accuracy enhancement project that will result in street center lines with an accuracy of 7.6 meters (source: *http://www.census.gov/geo/mod/backgrnd.html*). In rural areas, most of the TIGER files were initially derived from 1:100000 scale topographic maps, with some areas using other sources. You will recall that the spatial accuracy of such maps is plus or minus 167 feet.

The match between TIGER and the true location of features in rural areas is higher than in urban areas because the DLGs that were used as TIGER's source in these areas were originally derived from aerial photographs. Despite the reduced spatial accuracy, TIGER files still have tremendous value. TIGER files will typically contain more streets than do DLGs. In addition, TIGER will contain more detail about road systems, road names, and address ranges—which do not exist in DLGs—and will be more up-to-date than DLGs. Thus, for many applications (particularly those requiring address matching or local road detail) TIGER is preferred.

## CENSUS ADDRESS RANGES

An address range represents the possible valid addresses for a city block along a given side of a street. Typically, towns follow a plan for addressing where addresses begin at an origin point, and certain sides of streets contain either even or odd addresses. For example, in Denver, Colorado, addresses begin at the intersection of Broadway and Ellsworth. The first block of Broadway north of Ellsworth contains the address ranges of North Broadway from 0 to 98 on the even side, and from 1 to 99 on the odd side. In Denver, the east and south sides of streets contain even addresses, and the north and west sides have odd-numbered addresses.

For metropolitan areas, most of TIGER's streets have address ranges. If you think about how the Census used to be conducted (if you are old enough) prior to the mail-out/mail-back approach used today, census-taking required Census Bureau employees to conduct house-to-house interviews. Address ranges were important in the past for assignment of work for Census employees. An enumerator would need to know that he or she had to cover the 100 block of Maple Avenue, for example.

Street address ranges were therefore helpful in conducting past censuses. However, with the advent of TIGER and the geocoding of address questionnaires mailed back to the Census Bureau, address ranges became critical. However, TIGER coverage of address ranges is spotty in places, and in rural areas it may be nonexistent, despite the efforts by Emergency 911 and other services to address entire counties. Figure 6-1 presents a sample road layer showing that some street lines have names and address ranges, others have names only, and still others are completely blank. Efforts are underway to update the Master Address File, or MAF, so that TIGER will contain more and better address ranges. To read more about those efforts, go to *http://www.census.gov/geo/mod/maftiger.html*. Working with street data is discussed later in this chapter.

Fig. 6-1. Road attributes from TIGER. Note the missing values, partial Zip codes, and blank fields.

| LENGTH | FEDIRP | FENAME | FETYPE | FEDIRS | CFCC | FRADDL | TOADDL | FRADDR | TOADDR | ZIPL | ZIPR | CR |
|--------|--------|--------|--------|--------|------|--------|--------|--------|--------|------|------|-----|
| 0.0316 | | Pellissippi | Pky | | A31 | 0 | 0 | 0 | 0 | | | A |
| 0.01279 | | | | | A41 | 0 | 0 | 0 | 0 | | | A |
| 0.02399 | | Solway | Rd | | A41 | 0 | 0 | 3821 | 3817 | 37931 | | A |
| 0.21775 | | | | | A41 | 0 | 0 | 0 | 0 | | | A |
| 0.3709 | | Pellissippi | Pky | | A31 | 0 | 0 | 0 | 0 | | | A |
| 0.07085 | | | | | A63 | 0 | 0 | 0 | 0 | | | A |
| 0.19319 | | Oak Ridge | Hwy | | A31 | 8868 | 8800 | 8799 | 8741 | 37931 | 37931 | A |
| 0.17158 | | Dogwood | Rd | | A41 | 10899 | 10801 | 10898 | 10800 | 37931 | 37931 | A |
| 0.18031 | | Sam Crawford | Rd | | A41 | 3501 | 3517 | 3500 | 3514 | 37931 | 37931 | A |
| 0.10592 | | | | | A63 | 0 | 0 | 0 | 0 | | | A |
| 0.11098 | | | | | A41 | 0 | 0 | 0 | 0 | | | A |
| 0.05684 | | | | | A63 | 0 | 0 | 0 | 0 | | | A |
| 0.095 | | | | | A41 | 0 | 0 | 0 | 0 | | | A |
| 0.02219 | | Pellissippi | Pky | | A31 | 0 | 0 | 0 | 0 | | | A |
| 0.562 | | Pellissippi | Pky | | A31 | ✓ 0 | 0 | 0 | 0 | | | A |
| 0.08514 | | George Light | Rd | | A41 | 3598 | 3500 | 3599 | 3501 | 37931 | | A |
| 0.11275 | | George Light | Rd | | A41 | 3431 | 3499 | 3428 | 3498 | 37931 | | A |
| 0.61351 | | George Light | Rd | | A41 | 3426 | 3200 | 3425 | 3237 | 37931 | 37931 | A |
| 0.04748 | | | | | A41 | 0 | 0 | 0 | 0 | | | A |
| 0.12555 | | Oak Ridge | Hwy | | A31 | 8798 | 8732 | 8739 | 8737 | 37931 | 37931 | A |

Record: 4347   Show: All  Selected  Records (0 out of 28400 Selected.)   Options

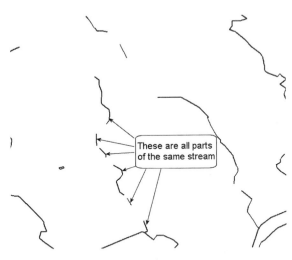

Fig. 6-2. Gaps in hydrography in TIGER.

# IMPROVING HYDROLOGY

Since TIGER was developed to support the Census, the inclusion and accuracy of hydrologic data was not a major concern. For example, it is not uncommon for streams not to connect where they pass through culverts. In figure 6-2, several sections of a stream (Third Creek in Knoxville, Tennessee) appear as disjointed water features. However, they are all part of the same stream. Starting with TIGER 2002, TIGER's water flag distinguished between perennial and intermittent water features. In future releases of TIGER, the Census Bureau plans to incorporate hydrologic information from the NHD.

# IMPROVING TOPOLOGY

TIGER enforces planar topology. In TIGER, every line segment has a starting and ending point, and wherever two lines cross a node is created. This can create problems when there are overpasses and underpasses. Slides 6 and 7 illustrate this point. Using Network Analyst in ArcView 3.2, a shortest path was calculated over the street layer for Knox County, Tennessee. The path includes a right-hand turn from Henley Street onto Neyland Drive. However, this would result in a vertical drop of about 100 feet!

Slides 6–7

Starting with TIGER 2002, fields were added to TIGER to account for overpasses and underpasses. As of this writing, those fields have not yet been populated, but future releases of TIGER should contain better information on overpasses and underpasses. Prior to the 2002 version of TIGER, all lines had unique TIGER line IDs, but nodes were not numbered. TIGER 2002 now includes Census Bureau assigned unique node ID numbers, called TZIDs, for all nodes in TIGER. These node IDs are necessary in implementing the overpass/underpass features.

# TIGER LINE LAYERS

In DLGs, various types of information are segregated into separate DLG files. There are nine such types of files (see Chapter 3 for details). TIGER takes a different approach. All street, boundary, hydrology, and other features are "vertically integrated" into one file. Polygons are constructed by assembling collections of lines. Thus, the most basic polygon in TIGER is the Census polygon, and each Census polygon has a Census ID field (CENID) and a polygon ID (POLYID).

Together, the CENID and POLYID give each basic polygon a unique identifier. The basic Census polygon may have no meaningful interpretation. However, collections of Census polygons may constitute meaningful entities, such as lakes, tracts, counties, and cities. Indeed, there are many possible types of features in a single TIGER archive. It is convenient to consider the many types of features by their topology—lines, polygons, and points.

Each line in TIGER has a Census Feature Classification Code, or CFCC. The CFCC codes consist of three characters (a letter and two numbers). From these, you can determine the type of feature the line represents. Table 6-2 presents the line features present in TIGER 2002 and their CFCC codes. More details about each layer can be found in the PowerPoint file for this chapter (see slides 8 through 10) and in Chapter 5 of the Census Bureau TIGER documentation.

Slides 8–10

**Table 6-2: TIGER Line Types**

| Layer Name | CFCC Code |
|---|---|
| Roads | A |
| Railroads | B |
| Miscellaneous ground transport | C |
| Landmarks | D |
| Physical features | E |
| Non-visible | F |
| Hydrography | H |
| Unknown | X |

In some versions of TIGER there are CFCC codes that start with P. These indicate provisional roads, which are those roads that were added "in preparation of Census 2000, but were not field verified . . . by census staff during field operations or through the use of aerial photography . . ." (TIGER Line Files UA 2000, page 3–41). Whether type A or P, the road layer is used for many purposes, including address matching and path analysis. Although an exhaustive explanation of each line type and CFCC code is beyond the scope of this chapter (and is available in the TIGER documentation), it is useful to look at the road layer in some detail.

> **NOTE:** *A complete list of CFCC codes and their meanings can be found in the file* CFCC.dbf *on the companion CD-ROM.*

Modern address-matching software is truly amazing. Think of all the street addresses in the United States. That is a lot of information. Yet, if you give most GIS software an address with a city and state name, it can search the entire United States TIGER-based database in less than a minute or two! If you give it a Zip code, it can do it in well less than a minute. Think about that: search the entire United States in less than a minute. The speed is based on clever ways in which software vendors index the data. If the Zip code is known, the software needs only to search a predetermined number of records, and those records will be indexed by Zip code.

Those records might be further indexed, probably on street name. If the software cannot find the address, it will most likely put it near where it thinks the address should be (on the correct street in the correct Zip code). If you have access to a large street database and geocoding software, you might wish to try an experiment. Search for an address using the Zip code. Then search for the same address using the city name and state. In most cases, using the street address and Zip code will be faster than using the street address and city name and state. This is because the indexing of streets on Zip codes will result in fewer records to search than by searching through an entire city.

TIGER uses a particular address format, commonly called US Streets, to store an address. If you think about addresses, you will soon realize that they are quite complex. They can contain apartment numbers, fractions (123 1/2 East Main Street), street names, prefix directions, suffix directions, and street types (e.g., Melrose Avenue and Melrose Place). Further, streets can be spelled (and misspelled) in a variety of ways. For example, suppose you wanted to look up an address on Twenty-Second Avenue. Would it be 22$^{nd}$ Ave, 22nd Ave, Twenty-second Ave, Twenty-Second Ave, or Twenty Second Ave? Are these all the same street? Furthermore, addresses for certain areas of the country do not follow the "standard" addressing practice. For example, "100 West 2200 South" is a common address form in the state of Utah.

Normally, a Zip code is associated with an address. Street addresses are usually associated with a point along a line, but a Zip code match is usually conceived of as a point in an area or polygon. Thus, there are two levels of geocoding: the polygon level (sometimes called the zone level) and the street level. Most address-matching software tries for a street address match (a point along a line), but if it cannot find one the software will give a Zip code match (a point in an area). TIGER has fields for streets that parse an address into various parts. These are listed in Table 6-3.

**Table 6-3: Components of Address Ranges**

| Field | Meaning |
|---|---|
| FEDIRP | Feature direction prefix (e.g., the first *S* in *S. Main St.*) |
| FENAME | Feature name (e.g., the *Main* in *S. Main St.*) |
| FETYPE | Feature type (e.g., the *St* in *S. Main St.*) |
| FEDIRS | Feature direction suffix (e.g., the *W* in *Maple St. W*) |
| FRADDL | From address left |
| TOADDL | To address left |
| FRADDR | From address right |
| TOADDR | To address right |
| ZIPL | Zip code left |
| ZIPR | Zip code right |

Some comments on these items are in order. For many streets, particularly in rural areas, these address fields will be blank. That is, just because TIGER contains fields for address attributes it does not mean that those fields are filled. Even in urban areas, these addresses fields may be only partially filled. For example, a road name

Slides 11–12

may be present but not its street address. A second point has to do with the left/right designation. The left- or right-hand side of a street is determined by the direction of digitizing. This means that a "from" address may be greater than the "to" address. It depends on in what direction the street was originally digitized. (See Power-Point slides 11 and 12.)

A final consideration is the Zip code information. You might think that by looking at the left- and right-hand Zip code fields you could construct closed polygons of Zip code areas. There are two reasons that this is not true. First, for many streets these fields are blank. That is, the Zip code information is often missing. A more fundamental reason, however, has to do with the nature of Zip codes. Zip codes are attributes of linear features. They are *not* area features. They never were designed to delineate polygons. TIGER does contain a polygon layer called Zip code tabulation areas, or ZCTAs, but these are approximations of Zip codes. There are no such things as "true" Zip code polygons in TIGER.

A more complicating aspect of street addresses is that they contain several fields with one-to-many relationships. Streets can have multiple names. For example, Main Street may also be State Highway 321, and U.S. Highway 12. That is, there can be alternate values of the fields FEDIRP, FENAME, FETYPE, and FEDIRS. Another type of one-to-many relationship has to do with address ranges. For some streets, there may be alternate addresses (including Zip code values) along a block. For example, if an office complex is built off of a road, that complex may have a different Zip code or street address range than was originally assigned to the street.

Finally, Zip + 4 information also may be present for street segments. A single street segment may have many Zip + 4 values on its left- and right-hand sides. In fact, a single building may have different Zip + 4 values for different floors. All of these possible one-to-many relationships present TIGER extraction software designers with a challenge when reporting the "true" street attributes. At least one software package you will study, TGR2SHP, reports all values present in the TIGER archive.

Slide 13

One final point needs to be made about TIGER addresses. You probably think of address ranges as numbers, such as 100 to 198 along Maple Street. In fact, many geocoding software packages have been created based on this assumption. However, there is at least one area in the United States (Queens, NY) where address ranges have non-numeric characters in them (see slide 13). For example, the From Address Left may be 68-01 and the To Address Left may be 68-09. The existence of a "-" (hyphen) in the address range can cause problems for some translation and geocoding software, resulting in fewer addresses geocoded, or some addresses geocoding to the wrong piece of geography.

# TIGER POLYGON LAYERS

There are many polygon layers present in TIGER, with the definitions of those layers dependent on the version of TIGER. For example, there are fields present in TIGER 1999 that are not present in TIGER 2002. It is also true that there may be fields in TIGER 2002 that are not in earlier versions of TIGER. Thus, each version of TIGER will have features that are unique to it, and each version may not contain features that are found in other versions of TIGER. However, some features (such as counties, tracts, block groups, and blocks) will be found in all versions of TIGER. Their boundaries and FIPSCODES may change.

It is convenient to break the TIGER polygon layers into those that are tied to Census Bureau enumeration units and those that are not. Those tied to Census enumeration units can be further divided into political areas and statistical areas. Political areas include any entity a government jurisdiction has control over, such as an Indian reservation, a city, or a county. Statistical areas are those defined by the Census Bureau for counting purposes. These include any areas that do not have legal jurisdiction but whose count is still reported, such as Census tracts, block groups, and Census designated places (communities). TIGER polygons that are not tied to Census counting include landmarks, water polygons, and key geographic locations. The enumeration unit polygons are further broken into versions, depending on the version of TIGER. For example, TIGER 2002 contains definitions of enumeration units for the year 2000 (the definition of boundaries for the 2000 Census) and for the current year (in this case, 2002).

The "current" definitions exist because area definitions change. Towns expand or contract, tract and county boundaries change, and so on. You might wonder which definition of an area you should use. It depends on the goal of your analysis. If your purpose is to show the current boundaries of an area, use the current geography. If you wish to use TIGER for thematic mapping of Census Bureau data products (such as the Economic Census or the 2000 Census of Population and Housing), use the definition of boundaries as they existed when the data were collected. Thus, to map data for Census 2000, use the 2000 polygon boundaries.

Because boundaries change, the TIGER files for one county may contain areas that were previously part of a neighboring county. That is, a given area may be in one county under the "current" geography but in another in

Slide 14

the 2000 geography. Thus, extracting 2000 county boundaries from TIGER 2002 may yield an incomplete delineation of a county, or may yield parts of two counties in a single county's TIGER data. An example of this phenomenon is presented in slide 14 of the PowerPoint file for this chapter.

Polygons that can be tied to the Census count data have FIPSCODE values. FIPS stands for Federal Information Processing Standard. These are usually numeric codes, although they may contain characters. However, they are not numbers. That is, you cannot add two FIPSCODEs together. The following example will illustrate how FIPSCODE values can be interpreted.

According to the Census Bureau, the census block in which the Burchfiel Geography Building (home of the Department of Geography at the University of Tennessee) falls has a block number of 2002. However, there are many blocks that have that number. They lie in different census tracts. In the case of the Department of Geography, the tract has a tract code of 000900. There may be other tracts within the state that share that code. The tract of interest here lies in Knox County, which has a county FIPS value of 093. There are several counties in the United States that have that FIPS value, but in this case we are only interested in the one in Tennessee. Tennessee's FIPS value is 47. Thus, the unique identifier for the block that contains the Department of Geography is the concatenation of the state, county, tract, and block FIPS values, or 470930009002002 (see figure 6-3).

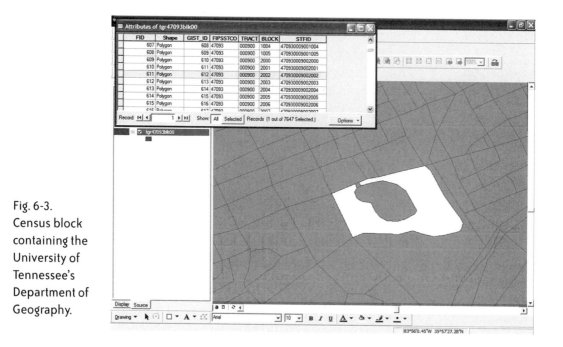

Fig. 6-3.
Census block
containing the
University of
Tennessee's
Department of
Geography.

All polygons for which the Census Bureau reports data collected in the Census of Population and Housing are built up from blocks. That is, blocks will always pack within other Census geography. This relationship is shown in the TIGER/Line Files 2002 Technical Documentation, which is reprinted here in figure 6-4. There are a few things to note from this figure. The categories Blocks, Block Groups, Tracts, Counties, States, and Nation all pack. As was illustrated previously, every block is within a tract, each tract is in a county, and each county is in a state. Although most layers pack within states, not all do. In particular, areas corresponding to American Indian lands, urban or metropolitan areas, and ZCTAs may cross state lines.

The unique identifier for any unit is often referred to as either the STFID or the SFID. The former (which was used through the 1990 Census) stands for Summary Tape File ID, whereas the latter (used in more recent Census documentation) stands for Summary File ID. The

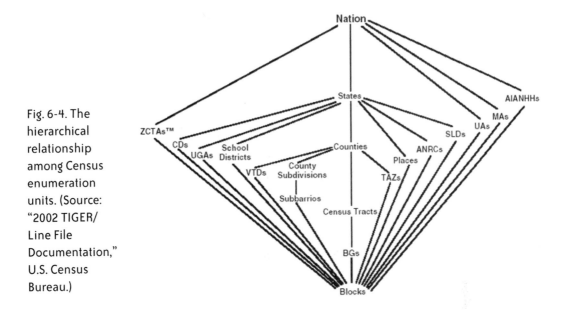

Fig. 6-4. The hierarchical relationship among Census enumeration units. (Source: "2002 TIGER/ Line File Documentation," U.S. Census Bureau.)

name change reflects the fact that technology has progressed to where tapes are seldom used anymore. The term *summary file* refers to the Census count files, discussed in Chapter 7. There are two important things to remember. You cannot determine the location of a block or tract simply by using its block or tract number. You must use the entire SFID, which includes the state and county FIPS values. The second important point is that this full FIPS code, or SFID, is the key value for relating census geography to census counts.

Polygons not directly tied to the census counts include landmark polygons, water polygons, and key geographic locations. Examples of landmarks are military bases, schools, and churches. Landmark and water polygons are related in that some, although not all, water polygons are landmarks. Further, landmark polygons can have one-to-many relationships with attribute tables. A landmark may have multiple names. For example, there are parts of the Bering Sea in Alaska that also are known as the Etolin Strait. In addition, each census polygon may be part of more than one landmark. For example, a polygon might be associated with a lake and a national park. Thus, a landmark (which might consist of many polygons) may have more than one name, or a polygon may be part of more than one landmark (each of which might have zero, one, or more names).

Key geographic locations (KGLs), such as shopping malls, may have more than one address. However, starting with TIGER 2002, KGLs were dropped from TIGER. More information on KGLs can be found in Census Bureau technical documentation and in the User Manual for TGR2SHP, discussed in material to follow.

The most basic polygons in TIGER are assigned CENID and POLYID identification. When combined, these give every topological polygon for TIGER in the entire nation a unique identifier. The CENID is the same for all polygons in a county, whereas the POLYID is a number assigned to each polygon. The polygons, much like ArcInfo coverages, start from number 2. Basic polygons may have no significance other than as topological entities. TIGER also contains centroid points for each of these polygons. Most users will not need to directly access the basic census polygons or their centroids, but they are in TIGER. Most often basic census polygons are used in redistricting applications. Polygons with more than topological significance (such as counties, tracts, blocks, places, and so on) can be viewed as being built up from combinations of one or more TIGER basic polygons (see slide 15).

Slide 15

The polygons that can be extracted from TIGER change with each version. Table 6-4 lists those supported by TIGER 2002. Not every layer will exist for every county. For example, Alaska native regional corporations will not be found outside Alaska, and sub-barrios will not be present outside Puerto Rico.

**Table 6-4: Polygons in TIGER 2002**

| Polygon Type | 2000 Geography | Current Geography |
|---|---|---|
| County | Yes | Yes |
| Tract | Yes | Yes |
| Block group | Yes | Yes |
| Block | Yes | Yes |
| American Indian/Alaska native/Hawaiian homelands | Yes | Yes |
| Alaska native regional corporations | Yes | Yes |
| American Indian tribal subdivisions | Yes | Yes |
| Consolidated city | Yes | Yes |
| County subdivision | Yes | Yes |
| Sub-barrio | Yes | Yes |
| Place | Yes | Yes |
| Elementary school districts | Yes | Yes |
| Secondary school districts | Yes | Yes |
| Unified school districts | Yes | Yes |
| Metropolitan statistical area/onsolidated metropolitan statistical areas | Yes | Yes |
| Primary metropolitan statistical areas | Yes | Yes |
| New England consolidated metropolitan areas | Yes | Yes |
| 106th congressional districts | Yes | Yes |
| Public-use microdata areas 1% sample | Yes | No |
| Public-use microdata areas 5% sample | Yes | No |

| Polygon Type | 2000 Geography | Current Geography |
|---|---|---|
| 5-Digit Zip code tabulation areas | Yes | Yes |
| 3-Digit Zip code tabulation areas | Yes | Yes |
| Urban 2000 | Yes | No |
| Traffic analysis zones | Yes | No |
| Voting districts | Yes | No |
| State legislative districts, upper chamber | Yes | No |
| State legislative districts, lower chamber | Yes | No |
| Urban growth areas | Yes | No |
| Landmark polygons | Yes | No |
| Water polygons | Yes | No |
| Corrected count polygons | Yes | No |

## Water Polygons

Water polygons present an interesting problem for users of TIGER. In particular, parts of water bodies are often included in the legal definition of a county. However, many people are unaccustomed to seeing a water body as part of a county. Consider, for example, figure 6-5. This is a portion of the *All Polygons* layer for Union County, New Jersey. The highlighted areas are coastal waters (Newark Bay). When we think of a county, we often visualize the county ending at the shoreline. However, a county's legal definition, and all its other census polygons, can extend well into surrounding water bodies. When creating thematic map of an area, you must decide whether or not you want to use the legal definition of the area or the land area definition.

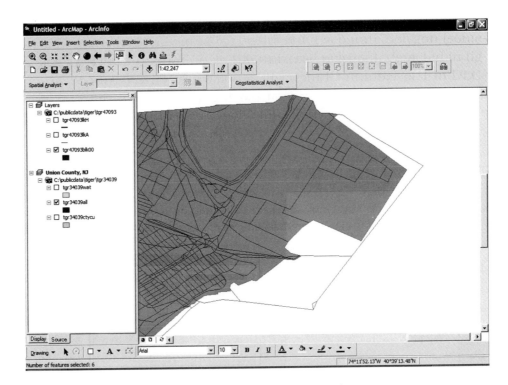

Fig. 6-5. Water polygons that are part of Union County, New Jersey.

Slide 16

An interesting example of this issue arose with the release of the CD108 TIGER. This version of TIGER contains the 108th congressional district boundaries. The Census Bureau depends on definitions of districts supplied by the states. Some states included all water polygons in the definitions of districts, and some did not. For example, the coastal waters of southern Connecticut were included in the 106th congressional district boundaries, but not in the 108th congressional district boundaries (see slide 16).

## ZCTAs

As discussed previously, there are no true Zip code polygons. However, TIGER files contain approximations of Zip code polygons called Zip code tabulation areas, or ZCTAs.

These are generalized versions of Zip code areas. ZCTAs were constructed to follow Census block boundaries. If more than one Zip code exists in a block, the ZCTA for that block will represent the Zip code for the majority of homes in that block at the time of the Census. There are two levels of ZCTAs: three- and five-digit. Three-digit ZCTAs are usually aggregations of five-digit ZCTAs. They fill all areas of the country but do not correspond to all Zip codes. That is, there are Zip code values that are not represented in the ZCTAs. This will occur for Zip codes that represent very few addresses. This is why, if you have ever address matched only on Zip codes, you may end up with some cases for which a corresponding ZCTA cannot be found.

ZCTAs have some properties of which you should be aware. They can cross county boundaries and in some cases state boundaries. Unlike counties, tracts, block groups, and blocks, they do not pack into higher-order Census polygons other than the nation. Further, most polygons in TIGER are defined by the Census Bureau or its partners to reflect various tabulation guidelines. ZCTAs are computer generated based on addresses within census blocks. For more information on how this is done, go to *http://www.census.gov/geo/ZCTA/zcta.html.*

There are two special five-digit ZCTA values. The first are codes that end in XX, as in 385XX. These correspond to three-digit ZCTAs that are not subdivided further into five-digit ZCTAs. The other special case contains codes that end in HH. These correspond to areas that are water features. Despite the fact that ZCTAs do not correspond precisely to postal Zip codes, they are still very useful for quickly locating features at scales where detailed street address matching is not required.

## Point Layers

Until recently, there were only a few explicit point layers in TIGER. A layer is explicit if TIGER attributes are associated with it. These layers consisted of the *All Polygon Centroid* layer mentioned previously, and of landmark points. Starting with TIGER 2002, more point information will be available from TIGER. Each topological point is now assigned a unique TIGER Node ID (TZID). Prior to that release, those nodes were implicit. That is, they existed but there were no attributes associated with them. In addition, future releases of TIGER will indicate if, in cases where more than two lines share a node, the node is an overpass or underpass for the lines it connects. The Census Bureau plans to incorporate better hydrography from the NHD data (see Chapter 8) in future TIGER releases. That, too, will increase the importance of point layers in TIGER.

## Data Files

TIGER uses its own distribution format, and the specifics of the format change with the version of TIGER. Thus, it is impossible to give a single, precise definition of what is inside every TIGER archive. Here is a brief overview of the content of some of the files that define the shape and topology of TIGER.

TIGER files, when unzipped, are ASCII (text) files. A complete set contains numerous files, with the exact number depending on the TIGER version. Each file contains a specific record type. Record type 1 contains information on lines. In particular, it will contain the TIGER Line ID (TLID), its starting and ending coordinates (in decimal degrees), the fields needed for street addressing, the CFCC value (which indicate if it is a road, rail, stream, and so on), and a set of attributes about the polygons on the left and right of each line. Example attributes are the

FIPSCODE of the state on the left and the right of the line, the county on the left and the right, and others.

Record type 2 contains the shape points for each line. The records in record type 2 (RT2) are matched to the lines in record type 1 by the TLID value. If a line does not have any records in RT2, it is a straight line segment. Record type I is the key to assembling lines that define polygons. Each record in RTI contains a TLID number and the CENID and POLYID of each basic census polygon on the left- and right-hand side of the line. Record types A and S contain information about the census enumeration units to which each basic census polygon belongs. The difference between RTA and RTS reflects the various dates (such as the Blocks Current and Blocks 2000 information categories) present in the TIGER archive.

The TLID value for a line is unique and never changes. If a line is split, each part is assigned a new TLID and the old TLID value is retired. Coincident lines, such as county boundaries, have different TLID values in each county's TIGER file.

To assemble a particular type of enumeration unit, say tracts 2000, one has to find all basic census polygons that belong to a given tract. Consider, for example, tract 000900 in Knox County, Tennessee. There will be a set of records in record type S that indicates the basic census polygons that belong to that tract. Once those basic census polygons are known, the lines that define them can be found from record type I. Each line that comprises a basic census polygon in the tract can then be considered. If the tract SFID on the left is equal to the tract SFID on the right, the line in question is not an external boundary to the tract and can be discarded.

If, on the other hand, this is not the case, the line in question is a boundary line for the tract and can be

Slide 17

added to the set of lines that define the tracts. Once all such lines are found, they can be combined so that they form a closed polygon that defines the tract (see slide 17). This may sound like a complicated process, and it is. However, there are software packages that do most of the tedious work for you. In addition to the record types discussed previously, there are several other record types that contain information on place names, legal description codes, and so on. These, too, have to be processed in order to extract all of the information in TIGER.

# OBTAINING TIGER DATA

As with the data sources covered in the previous chapters, map libraries and state user groups are good sources of TIGER files. Sometimes the TIGER files have already been processed into usable map layers, as in TIGER 2000 on the Geography Network (*http://www.geographynetwork.com* and *http://www.esri.com/data/download/census2000_tigerline/*). These shape files were processed using TGR2SHP, written by this author. However, preprocessed TIGER files may not have all of the additional one-to-many data tables that can be generated from TIGER, or they may not have the most up-to-date TIGER version. From the site, the user can select counties, specific layers of TIGER data within each county, and Census population and housing data for the area needed.

The main source of TIGER files is the U.S. Census Bureau's TIGER web page at *http://www.census.gov/geo/www/tiger/index.html* (see figure 6-6). You can download various versions of TIGER directly from links at this site. If you use an ftp program, the anonymous ftp site is *ftp2.census.gov/geo/tiger*. The TIGER files are organized by state and come as a zipped archive of files. (See the PowerPoint file, slides 18 through 21, for details.)

Slides 18–21

When downloading the files, be sure to note the version of TIGER you are obtaining.

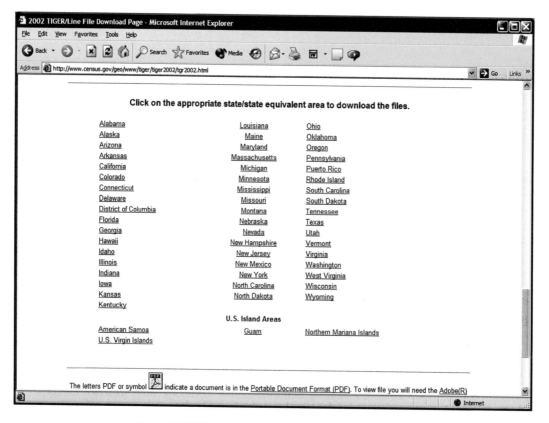

Fig. 6-6. TIGER download web page.

As mentioned at the beginning of the chapter, there are many versions of TIGER. Prior to TIGER 97, each TIGER version has a specific file extension that indicated its version number. For example, TIGER 95 files all had extensions that ended in *.f6\**, such as *.f61*, *.f62*, and so on. Since TIGER 97, all TIGER files use the same extension template, *.rt\**. Consequently, any TIGER version released since that time contains files with names and extensions identical to those in any version of TIGER since TIGER 97.

Put simply, you cannot tell the TIGER version by looking at the file name. There is a date in each record type that indicates when the file has been processed, but there have been at least two versions of TIGER that had overlapping processing dates. Without knowing the version of TIGER you are using, you may not be able properly extract the information inside the TIGER archive.

# EXTRACTION METHODS

Two extraction methods are covered here: TGR2SHP—a standalone program for extracting TIGER layers—and a TIGER extraction tool from ESRI. TGR2SHP handles versions of TIGER from TIGER 94 to TIGER 2003. The ESRI toolbox contains a tool for TIGER 1995 to 2000.

## TGR2SHP

TGR2SHP allows you to choose the TIGER layers you wish to extract. It extracts the one-to-many information links described previously and does so relatively quickly and easily. It also allows you to process several counties at once and to merge them into single themed shapes, such as all roads, all counties, and so on. You can clip water polygons if you wish to do so. In the interest of full disclosure, I should point out that I wrote this program, and therefore my opinions are not unbiased! The setup program TGR2SHP can be found on the companion CD-ROM, along with the User Manual.

To use TGR2SHP, simply start the program and choose the TIGER archives you wish to process. You do not have to unzip the TIGER archives (although you can). The program then allows you to select the layers you wish to extract (see figure 6-7). Once completed, the program creates shapefiles for the layers you requested. There is a

MapInfo Interchange File version, called TGR2MIF, available as well.

**Tiger 2002 Extraction Options**   ☒

Controls

| Select All | Clear All | ☐ Clip Perennial Water Polygons | ☐ Use these options for all subsequent cases |

☐ Roads

☐ Rails

☐ Misc. Ground Transport

☐ Landmarks

☐ Physical Features

☐ Non-Visible

☐ Hydrography

☐ Unknown

☐ County 2000

☐ County Current

☐ Tract 2000

☐ Tract Current

☐ Group 2000

☐ Group Current

☐ Block 2000

☐ Block Current

☐ American Indian/ Alaska Native/ Hawaiian Homeland 2000

☐ AIANHH Curren

☐ Alaska Native Regional Corporations 2000

☐ ANRC Current

☐ American Indian Tribal Subdivisions 2000

☐ AITS Current

☐ Consolidated City 2000

☐ Consolidated City Current

☐ County Subdivision 2000

☐ County Subdivision Current

☐ Subbarrio 2000

☐ Subbarrio Current

☐ Place 2000

☐ Place Current

☐ School Districts Elementary 2000

☐ SDELM Current

☐ School Districts Secondary 2000

☐ SDSEC Current

☐ School Districts Unified 2000

☐ SDUNI Current

☐ MSA/CMSA 2000

☐ MSA/CMSA Current

☐ PMSA 2000

☐ PMSA Current

☐ NECMA 2000

☐ NECMA Current

☐ 106th Congression Districts

☐ Congressional Districts Current

☐ PUMA 5%

☐ PUMA 1%

☐ ZCTA5 2000

☐ ZCTA5 Current

☐ ZCTA3 2000

☐ ZCTA3 Current

☐ Urban 2000

☐ Traffic Analysis Zones

☐ Voting Districts

☐ State Legislative Districts, Upper House

☐ State Legislative Districts, Lower House

☐ Urban Growth Area

☐ Landmark Pts and Polys

☐ Water Polygons

☐ Corrected Count Polys

☐ All Nodes, Lines, Polys and Centroids

OK

Fig. 6-7. Layer extraction options in TGR2SHP for TIGER 2002.

TGR2SHP allows you to organize the resulting shapes by county, by theme, and by theme-and-merge. You also can write all shapes to a single directory (see figure 6-8). The merge option allows you to process all counties in a state, say, to produce a state map of each TIGER layer.

Slides 22–23

As with DLG2SHP, TGR2SHP gives you all information necessary to support the one-to-many relationships between spatial features and attributes. Slides 22 and 23 illustrate the one-to-many relationship. TGR2SHP also creates some fields that are helpful in thematic mapping. For example, the SFID is created for polygons that can be joined to demographic data from the Census 2000, and two-character CFCC codes are created to facilitate mapping of roads by road class. A detailed description of the program can be found in the TGR2SHP manual, *Manual.pdf*, on the companion CD-ROM.

**TGR2SHP Output Options**

Output Parent Directory: D:\ct      [Browse]

There are four ways to organize your output files:

    Group shapes by county (a directory for each county's shapes)
    Put all shapes in a single directory
    Group shapes by theme (a directory for each type of shape, eg,roads, tracts, etc)
    Group shapes by theme and merge (a single shape for each map layer)

Please choose an option

**Output Organization Options -- Choose One**

    ● Organize by County - Create a directory for each county

    ○ Put all outputs in a single directory

    ○ Organize by Theme - Create a directory for each map layer

    ○ Organize by Theme and Merge - For each layer create a single shape

**Address Ranges Field Types**

    ● Address Ranges as Numbers      ○ Address Ranges as Characters

Some counties have characters in their address ranges. However some geocoding software requires address ranges to be numbers. Choose the option you wish to use.

[ OK ]

Fig. 6-8. Output options in TGR2SHP.

## Tiger to Coverage Wizard

ArcToolbox contains a wizard that allows you to import certain versions of TIGER to ArcInfo coverages. The wizard, along with its ArcWorkstation companion TIGERTOOL, creates one coverage of line and polygon topology and one coverage of point topology. The wizard is limited to versions of TIGER from 1995 to 2000. An example of working with the Import tool can be seen in slides 24 and 25.

Slides 24–25

While reading about basic census polygons, you might have realized that it is possible to import just the basic polygons and then dissolve them to the level you need.

For example, you might use your GIS software to dissolve basic census polygons into tracts. This is the approach the ESRI tools take. They produce a polygon coverage of basic census polygons only, along with attributes that can be used to determine which polygons can be combined to create specific layers. This polygon layer is similar to the All Census Polygons shape created by TGR2SHP.

The following is an example of how to create tracts from the polygon coverage created by the TIGER to Coverage wizard. Using the GeoProcessing wizard in ArcMap, select the option for dissolving the polygons based on their tract value (see figure 6-9). You will be given the chance to choose other attributes (such as the county FIPSCODE) to include in the resulting layer. The resulting layer is of tracts for that county.

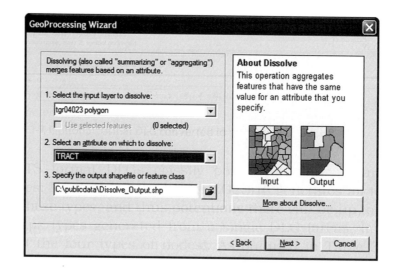

Fig. 6-9. Dissolving on tracts.

This may seem relatively easy, but there are potential pitfalls with this approach. There are two points to keep in mind when working with the ESRI-generated polygons. Earlier in this chapter you learned that complete FIPSCODES, or SFIDs, were combinations of several levels of FIPSCODES. For example, a unique tract identifier is a combination of the state, county, and tract FIPSCODE values. The previous dissolve example worked because only one county was being processed. Consider what would happen, however, if you tried the same approach with blocks or block groups.

The GeoProcessing wizard in ArcGIS only allows single field dissolves. Because block values can repeat from tract to tract, and tracts from county to county, dissolving based on the block value will not give the correct result. In addition, the TIGER to Coverage wizard in Arc-Toolbox treats FIPSCODES as numbers, not strings. (In ESRI's defense, the Census Bureau documentation in places refers to FIPSCODE values as numbers. They are only numbers in the sense that Zip codes are numbers. You cannot manipulate them mathematically.) The result is that before creating a new field consisting of state, county, tract, and block fields, you must convert these values to strings before concatenating them.

Moreover, you must add any missing leading zeros to those fields. Using the example of the block that contains the Department of Geography at the University of Tennessee discussed previously, concatenating fields generated by the ESRI extraction tool would yield a FIPSCODE of 47939002002, when it should be 47**093000**9002002. (The missing leading zeros on the tract are in bold.) Although the first number may be unique, it will not match the Summary File fields from which you might draw attribute data. It also will be difficult, if not impossible, to extract the state and county from such a code.

The lines associated with this coverage contain all types of lines in TIGER. That is, all of the line types in Table 6-1 are in a single coverage. You will have to select and save those for roads, rails, hydrology, and other layers separately within your GIS. The Import tool will not do that for you.

In addition to the coverages created with their attributes, the Import wizard also creates an Info table for each record type (see figure 6-10). You can then join or link the appropriate record types with their corresponding features. For example, you could create a one-to-may relationship between the line features and the Zip + 4 codes found in record type Z. These tables can be related by the TLID value.

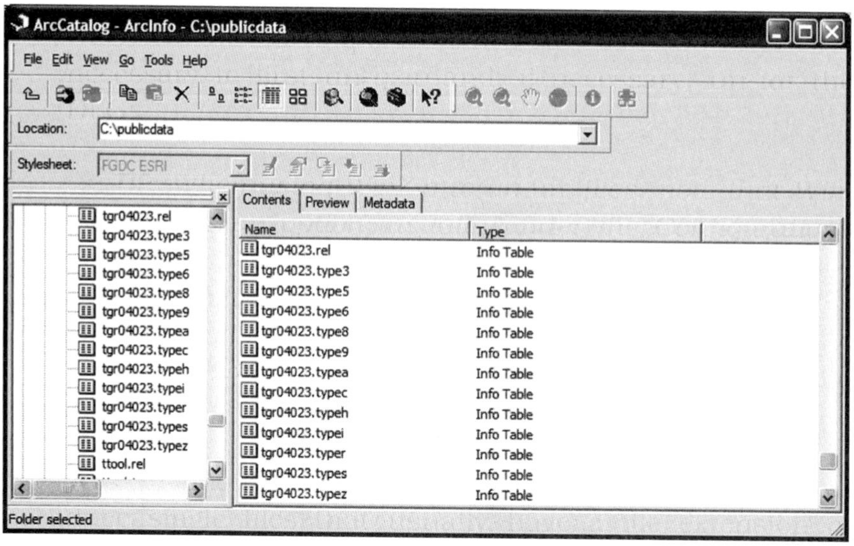

Fig. 6-10. Info tables for each record type, plus the relate table (*.rel*).

Slide 26

To make full use of the Info tables, you will need to carefully study the relationships among the various features and the various record types (see slide 26). Some of these require the use of multiple keys. For example, record type C contains information on names. To associate the proper name with a polygon may require matching the data year (which is not a field reported in the polygon attribute table), FIPSCODE, and possibly other fields, such as the LSADC (Legal/Statistical Area Description Code).

Making matters more complicated, building the proper relationships between features and attributes may require tracing linkages through several files. For example, a polygon that is part of a landmark will have a record in the basic census polygon coverage (indicating its CENID and POLYID) and in record type 8. That record will contain a landmark ID. The landmark ID is also in record type 7. It is from that record you can find the CFCC value and landmark name. In addition, record type C may have additional names for the polygon. The relationships among the various record types are depicted in figure 6-11, from the Census Bureau.

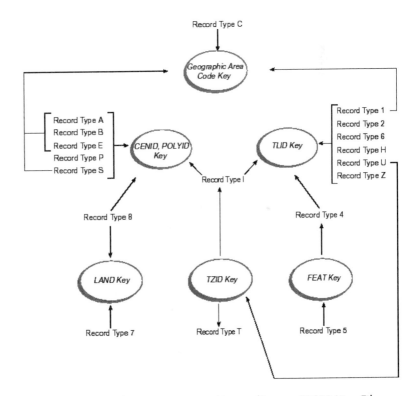

Fig. 6-11. Relationships among record types. (Source: TIGER Line Files Technical Documentation 2002.)

# TIGER AND GEODATABASE MODELS

As this text goes to press, ESRI is releasing a data model for TIGER. The data model from ESRI consists of a personal geodatabase (a Microsoft Access file) that contains editing rules required for TIGER updates to be valid. That is, the database is seeded with the rules inherent in figure 6-4.

TIGER is a collaborative database. That is, it is maintained by various local agencies, such as metropolitan planning organizations, state planning offices, and the like. It remains to be seen if this data model will become widely used by the communities who build and maintain TIGER. It is unlikely the data model will become widely adopted by the TIGER user community. The topological rules on the relationships between various polygon levels in TIGER require a certain amount of computational overhead that makes the capability of limited value to those users who simply want to make a thematic map of census data or geocode an address. For those users, shapefiles are the appropriate level of complexity.

Whatever method you use to extract information from TIGER, it is likely that you will wish to use the enumeration areas you derive to make thematic maps of population and housing based on data collected by the Census Bureau. Much of that data is distributed in summary files, the topic of the next chapter.

# TIGER 2003 AND THE 2002 ECONOMIC CENSUS

The Census Bureau plans to release TIGER 2003 during the first quarter of 2004. The main purpose of this version of TIGER is to support the 2002 Economic Census.

That census will be released starting in 2004 and continuing through 2006. The schedule for reports from the 2002 Economic Census can be found at *http://www. census.gov/epcd/www/guide.html.* TIGER 2003 will contain new metropolitan area definitions developed for the 2002 Economic Census. A list of these new polygons can be found at *http://www.census.gov/population/www/ estimates/metrod.html.* The full documentation for this new version of TIGER is yet to be released. However, the following describes some of the features that will be new in this version of TIGER.

Slide 27

Several new polygons will be used in the 2002 Economic Census (see slide 27). These polygons are delineated in such a way as to make it possible to find economic core zones and their surrounding economically integrated areas. Such areas are called core-based statistical areas (CBSAs). The economically integrated areas are based on commuting ties. CBSAs usually consist of a central county or counties and those surrounding counties that meet one or both of the following criteria.

- At least 25 percent of the surrounding county's residents work in the CBSA central county or counties

- At least 25 percent of the jobs in the surrounding county are accounted for by workers in the CBSA's central county or counties

The percentages represented in these rules are used to define "employment interchange." The employment interchange is defined as "the sum of the percentage of commuting from the entity with the smaller total population to entity with the larger population and the percentage of the employment in the entity with the smaller total population accounted for by workers residing in the entity with the larger total population" (source: *Federal Register,* vol. 65, no. 249, p. 82234). The second rule reflects the phenomenon of reverse commuting.

CBSAs must contain at least one metropolitan statistical area (MeSA) or micropolitan statistical area (MiSA). MeSAs have at least one urbanized area with a population of 50,000 or more. MiSAs must have one or more urban clusters with a population of at least 10,000 people but less than 50,000 per cluster. Together, MeSAs and MiSAs account for 93 percent of the U.S. population. Because CBSAs can be very large, they are sometimes disaggregated into metropolitan divisions (METDIVs). METDIV areas are derived from CBSAs with a population of 2.5 million or more in a single core. In addition to disaggregation, adjacent MeSAs and MiSAs may be aggregated to form combined statistical areas (CSAs).

Adjacent MeSAs and MiSAs are combined automatically if their employment interchange exceeds 25 percent. Areas with at least 15- but less than 25-percent economic interchange may be combined to form CSAs. CSAs are similar to the traditional CMSAs (combined metropolitan statistical areas) used in previous versions of TIGER. For most of the country, the basic units for defining CBSAs, CSAs, MeSAs, MiSAs, and METDIVs are counties. That is, these economic census polygons are built up from counties. However, New England is treated differently. In that region, the basic units for defining these areas are cities and towns (see slides 28 and 29).

Slides 28–29

In addition to containing the area definitions of the new polygons, definitions of states, counties, and places that apply to the 2002 Economic Census are also included in TIGER 2003. Although only few state and county definitions will change from 2000, most place definitions will change. That is, place definitions used for the 2000 Census of Population and Housing will not match the 2002 Economic Census delineation of places.

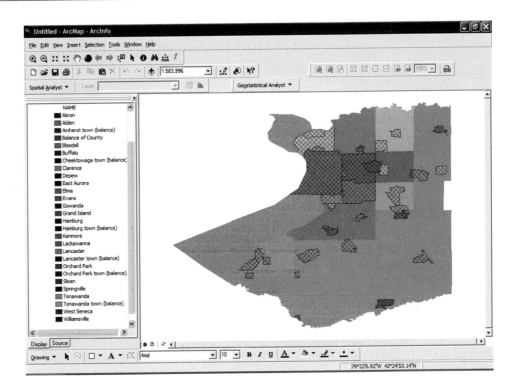

Fig. 6-12. Comparing place definitions for Erie County, New York.

Figure 6-12 presents an example from Erie County, New York. In this figure, the solid shaded areas represent places defined for the 2002 Economic Census. The hatched areas are places as defined for the 2000 Census of Population and Housing. The Economic Census places fill the entire county, whereas the 2000 places do not. In fact, for the Economic Census, a new place type called Balance of County is used to make sure the entire county is assigned to a place. Balance of County polygons do not need to be contiguous and may include major water bodies (see slide 30).

Slide 30

The version of TGR2SHP on the companion CD-ROM will work with TIGER 2003. The polygons specific to the Economic Census appear at the bottom of the output options dialog (see figure 6-13). The resulting CBSA or NECTA shapes will contain both MeSAs and MiSAs. To differentiate between them, use the Legal Statistical Area Description Code (LSADC). Table 6-5 lists the LSADC codes (and their meanings) that are new in TIGER 2003.

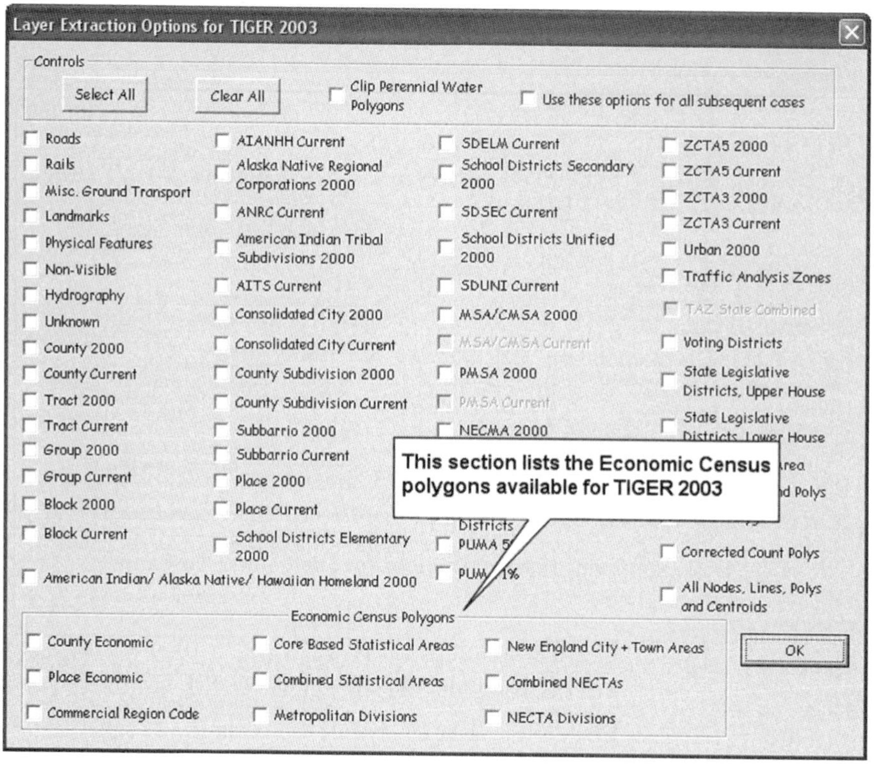

Fig. 6-13. The Economic Census polygon options in TGR2SHP 2003.

## Table 6-5: LSADC Codes for 2002 Economic Census Polygons

| LSADC | Meaning |
|-------|---------|
| M0 | Combined Statistical Area |
| M1 | Metropolitan Statistical Area in CBSA |
| M2 | Micropolitan Statistical Area in CBSA |

| LSADC | Meaning |
|-------|---------|
| M3 | Metropolitan Division |
| M4 | Combined New England City and Town Area |
| M5 | New England City and Town Area Metropolitan Statistical Area |
| M6 | New England City and Town Area Micropolitan Statistical Area |
| M7 | New England City and Town Area Division |

# WEB SITES AND DOCUMENTATION

TIGER documentation is available as PDF files from the Census Bureau. Each version of TIGER will have its own technical documentation, and these can be found on the TIGER web site (*http://www.census.gov/geo/www/tiger/*). The TIGER documentation has some excellent graphics for defining topology, address range and Zip code issues, and the relationships between polygon types.

The Geography Division of the Census Bureau also has a web site with the title "Understanding Census Bureau Geography" (*http://www.census.gov/geo/www/reference.html*). This site contains links to web pages and technical documents that explain various Census geography criteria and definitions. For example, there are links to pages explaining the relationship between Zip codes and ZCTAs, metropolitan area definitions, American Indian/Alaska Native/Hawaiian Homeland polygons, and the uses of FIPS codes. The Geography Division has its own home page (*http://www.census.gov/geo/www*), which contains links to its various products and programs.

# CHAPTER 7

# CENSUS SUMMARY FILES

## INTRODUCTION

A COMMON USE OF GIS IS TO MAKE THEMATIC maps of demographic information. TIGER files supply the geographic units and linkages needed to construct such maps. However, the data about population and housing collected by the Census Bureau are not included in TIGER. For that information, you need to use information distributed in other file sets. The attribute data collected by the Census Bureau is distributed in various forms, two of which are the focus of this chapter: Summary File 1 (SF1) and Summary File 3 (SF3).

When the 2000 Census of Population and Housing was taken, most households were sent a brief survey called the Short Form. This form asked questions about population and housing. Population questions requested information on numbers of people, race, ethnicity, gender, and age. Household information collected included information on occupancy status, ownership or renter status, and the like. A copy of the Short Form questionnaire can be found in Appendix D of the Summary File 1 technical documentation at *http://www.census.gov/prod/cen2000/doc/sf1.pdf.*

Not every household was asked to fill out the Short Form. Approximately 1 out of every 6 households was asked to complete the Long Form. The Long Form contained questions found on the Short Form. It also contained other questions, including those about place of birth, education, income, migration, disability status, commuting time, automobile ownership, and many other variables. A copy of the Long Form questionnaire can be found in Appendix D of the Summary File 3 documentation at *http://www.census.gov/prod/cen2000/doc/sf3.pdf*. Estimates for the entire population were generated from this sample of approximately 19 million households.

The two file types you will study in this chapter correspond to these two types of forms. SF1 contains the 100-percent sample results for the questions asked of all people and households. SF3 contains the estimates for the entire population based on the sample of 19 million households (see slides 1 and 2). Because the results in SF3 are based on a statistical extrapolation, they will not necessarily match those found in SF1. For example, for the same geographic area the number of people who are Hispanic may differ in SF1and SF3. Table 7-1 presents an overview of the summary files.

Slides 1–2

## Table 7-1: An Overview of Census 2000 Summary Files

| Item | Summary File 1 | Summary File 3 |
|------|----------------|----------------|
| Data source | 100-percent sample of questions found on Short Form. | Extrapolated results from Long Form sample. |
| Data content | Information on population and housing, including numbers by race, ethnicity, age, occupancy status, and so on. | More detailed socio-demographic information, including place of birth, education, income, commuting times, and automobile ownership. |

| Item | Summary File 1 | Summary File 3 |
|---|---|---|
| Data size | 286 tables containing over 8,000 variables. | 816 tables and over 16,500 variables. |
| Data distribution | 40 zipped files containing data for all summary levels. | 77 zipped files containing data for all summary levels. |
| Mapping considerations | Must match FIPS and Census codes to geographic boundaries from TIGER. | Must match FIPS and Census codes to geographic boundaries from TIGER. |
| Software considerations | For extracting a few variables for a few places, use the Census web site. Otherwise, use extraction software or database methods. | For extracting a few variables for a few places, use the Census web site. Otherwise, use extraction software or database methods. |
| Other considerations | The key file for connecting table values to TIGER geography is the Geographic Header File. Some summary levels have geographic components. | The key file for connecting table values to TIGER geography is the Geographic Header File. Some summary levels have geographic components. |
| Important consideration | There is a significant change in how racial and ethnic data are reported in the 2000 Census. In particular, multiple racial and ethnic responses are permissible. This makes comparisons to earlier census years difficult. | There is a significant change in how racial and ethnic data are reported in the 2000 Census. In particular, multiple racial and ethnic responses are permissible. This makes comparisons to earlier census years difficult. |
| Software used | SF1toTable, Microsoft Access, the American FactFinder web pages. | SF3toTable, Microsoft Access, the American FactFinder web pages. |

SF1 and SF3 contain information for Census Bureau enumeration units, such as blocks, block groups, and Census tracts. This information is distributed in a set of files that contains data on each type of enumeration unit. The file sets are distributed at two levels of aggregation: state summary files (one set of files per state or state equivalent) and national files. The keys to understanding how data are reported are the Summary Levels, the Geographic Component values, and the structure of what is called the Geographic Header File.

# SUMMARY LEVELS

As you saw in the discussion of TIGER, there are many levels of enumeration units, such as blocks, block groups, and Census tracts. These were depicted in figure 6-4. Each tract, for example, has a unique FIPSCODE value, constructed by concatenating the state, county, and tract values. In the summary files, a summary level value consists of a three-digit code, and there is one code for each type of enumeration unit. Consider, for example, tracts. The code for tracts is 080. Thus, any summary file record that has a corresponding summary level of 080 represents a tract. A complete list of summary levels can be found in Chapter 4 of the Summary File 1 Technical Documentation (*sf1.pdf*) and the Summary File 3 Technical Documentation (*sf3.pdf*).

The summary level values present in a set of summary files will depend on whether the set represents a state or the nation. For example, tracts (summary level 080) exist in the state summary files, but are not present in the national files. To obtain tract values for all tracts in the United States, you have to combine the results from each state's summary files. The number of summary levels also varies by file type. For state level SF1 files there are 61 summary levels, whereas for state level SF3 there are 68.

Each summary level corresponds to a level of enumeration unit. In addition, some levels correspond to subdivisions of units. For example, summary level 871 refers to ZCTAs (Zip code tabulation areas) in state-level files. Summary level 881 indicates ZTCA values by county. These summary levels exist only for state summary files. Hence, summary level 871 yields values of ZCTA by state. In the national-level file sets, summary level 860 yields the values of the ZCTA over all states. That is, if a ZCTA crosses state boundaries (see the discussion of ZCTAs in Chapter 6) summary level 860 corresponds to ZCTA-level counts, 871 corresponds to ZCTA by state-level counts, and 881 corresponds to ZCTA by county-level counts.

Figures 7-1 and 7-2 present the SF3 state summary options. These are screen shots of extraction software (*SF3toTable*) discussed in material to follow. A complete list of summary levels from this summary file extraction software can be found in slide 3 of the PowerPoint file for this chapter.

Slide 3

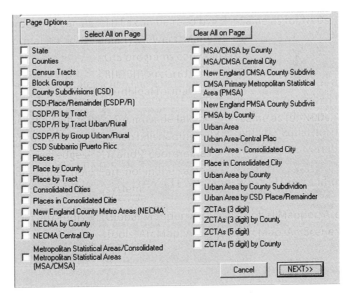

Fig. 7-1. State summary levels for SF3, Part 1.

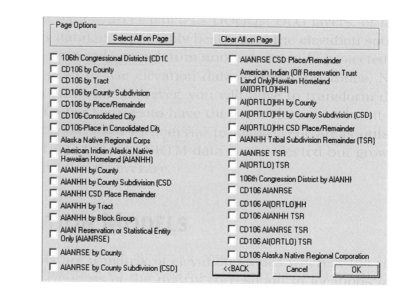

Fig. 7-2. State summary levels for SF3, Part 2.

# GEOGRAPHIC COMPONENTS

In addition to reporting information by summary level, the Census Bureau also reports information on various geographic components for selected summary levels. You might think that any table for a state extracted from the state-level files would have one entry only: the values of census variables for that state. This is not the case. For example, in addition to reporting the population of a state, state summary files for SF1 contain the population of the state in MSA/CMSAs; in MSA/CMSAs of various sizes, such as more than 5,000,000; in Tribal Designated Statistical Areas; and so on.

Slides 4–5

There are up to 98 possible geographic components. However, some components will only apply in specific cases (see slides 4 and 5) For example, Alaska Native Village Statistical Areas will not be found outside Alaska. Similarly, components that refer to New England County Metropolitan Areas will not be found outside New England. A complete list of the geographic components used in SF1 and SF3 can be found in Footnote 3 of the Footnotes section of each file's technical documentation (*sf1.pdf* and *sf3.pdf*).

A table of summary levels that have geographic components can be found in Chapter 4, Summary Level Sequence Chart, of the technical documentation. Table 7-2 lists the possible geographic components at the state level and their total populations for the state of Delaware. A consequence of this style of reporting is that TIGER polygon layers may have a one-to-many relationship with tables in the SF1 and SF3. This will only be the case for those summary levels that contain geographic components.

## Table 7-2: Geographic Components for Delaware at the State Level for SF1

| Geographic Component | Population |
|---|---|
| Entire state | 783600 |
| In MSA/CMSA | 626926 |
| In MSA/CMSA of 5,000.000 or more population | 500265 |
| In MSA/CMSA of 2,500.000 to 4,999.999 population | 0 |
| In MSA/CMSA of 1,000.000 to 2,499.999 population | 0 |
| In MSA/CMSA of 500.000 to 999.999 population | 0 |
| In MSA/CMSA of 250.000 to 499.999 population | 0 |
| In MSA/CMSA of 100.000 to 249.999 population | 126697 |
| In MSA/CMSA of 50.000 to 99.999 population | 0 |
| In MSA/CMSA in MSA/CMS central city | 133346 |
| In MSA/CMSA of 5,000.000 or more population (in MSA/CMSA central city) | 101211 |

| Geographic Component | Population |
|---|---|
| In MSA/CMSA of 2,500.000 to 4,999.999 population (in MSA/CMSA central city) | 0 |
| In MSA/CMSA of 1,000.000 to 2,499.999 population (in MSA/CMSA central city) | 0 |
| In MSA/CMSA of 500.000 to 999.999 population (in MSA/CMSA central city) | 0 |
| In MSA/CMSA of 250.000 to 499.999 population (in MSA/CMSA central city) | 0 |
| In MSA/CMSA of 100.000 to 249.999 population (in MSA/CMSA central city) | 32135 |
| In MSA/CMSA of 50.000 to 99.999 population (in MSA/CMSA central city) | 0 |
| In MSA/CMSA (not in MSA/CMS central city) | 493616 |
| In MSA/CMSA of 5,000.000 or more population (not in MSA/CMSA central city) | 399054 |
| In MSA/CMSA of 2,500.000 to 4,999.999 population (not in MSA/CMSA central city) | 0 |
| In MSA/CMSA of 1,000.000 to 2,499.999 population (not in MSA/CMSA central city) | 0 |
| In MSA/CMSA of 500.000 to 999.999 population (not in MSA/CMSA central city) | 0 |
| In MSA/CMSA of 250.000 to 499.999 population (not in MSA/CMSA central city) | 0 |
| In MSA/CMSA of 100.000 to 249.999 population (not in MSA/CMSA central city) | 94562 |
| In MSA/CMSA of 50.000 to 99.999 population (not in MSA/CMSA central city) | 0 |
| Not in MSA/CMSA | 156638 |

| Geographic Component | Population |
|---|:---:|
| American Indian reservation and trust land (Federal Tribe) | 0 |
| American Indian reservation and trust land (State Tribe) | 0 |
| Oklahoma Tribal Statistical Area | 0 |
| Tribal Designated Statistical Area | 0 |
| Alaskan Native Village Statistical Area | 0 |
| State Designated American Indian Statistical Area | 22683 |
| Hawaiian Home Land | 0 |

# CENSUS TABLES AND VARIABLES

SF1 and SF3 generate many tables and variables. SF1 contains 286 tables and over 8,000 variables, whereas SF3 contains 816 tables with over 16,500 variables. The tables in both files can be categorized by the types of variables they contain, lowest level at which those variables are reported, and if information is further reported by race. Both SF1 and SF3 contain tables that pertain to population and housing variables.

For SF1, the lowest level of reporting is the block level. Population variables reported at the block level are in tables that begin with the letter P, and those for housing begin with the letter H. For example, table P09 is Population Table 9. It will contain variables at the block level or higher. For certain population variables, reporting at the block level would compromise confidentiality. Those variables are reported at the tract level or above. Population tables at that level have the prefix PCT.

In addition to organizing variables by type and level (P, H, PCT), the Census Bureau reports variables by major race and Hispanic or Latino group. Such tables are given a one-letter suffix. The suffix indicates the group to which the table applies. For example, table H15H is a housing table that reports data for Hispanic or Latino householders. Table 7-3 lists the table suffixes and the racial or ethnic groups to which they apply.

**Table 7-3: Major Racial/Ethnic Classifications and Corresponding Suffix Letters**

| Suffix Letter | Racial or Ethnic Group |
|---|---|
| A | White alone |
| B | Black or African-American alone |
| C | American Indian or Alaskan Native alone |
| D | Asian alone |
| E | Native Hawaiian or other Pacific Islander alone |
| F | Some other race alone |
| G | Two or more races |
| H | Hispanic or Latino |
| I | White alone, not Hispanic or Latino |

A similar tabular naming strategy is used in SF3. Because much of the information in the Long Form is considered more sensitive than that in the Short Form, the lowest level of reporting for SF3 is the block group. Tables beginning with P indicate population, and those with H indicate housing. PCT indicates population tables at the tract level or above, whereas HCT indicates housing variables reported at the tract level or above. The major racial group and Hispanic or Latino suffixes in Table 7-3 are used for these tables as well.

Keeping track of all of the tables and variables can be a daunting task. There are some things to aid in your search for the right variables and tables. Your author has created a series of hyperlinked files that contain file titles (see figure 7-3). These can be found on the companion CD-ROM. By clicking on a file title, you can find the variable definitions for the content of that file. For example, figure 7-4 illustrates the results of clicking on Table P.6, Race. Files relating to SF1 are named *sf1\*.html*, whereas those relating to SF3 are named *sf3\*.html*. In addition to these files, the Census Bureau has created files that contain the same information but in DBF format. There is one file for SF1 and one for SF3. Some extraction software (such as *SF1toTable* and *SF3toTable*, which can be found on the companion CD-ROM) will list tables by name.

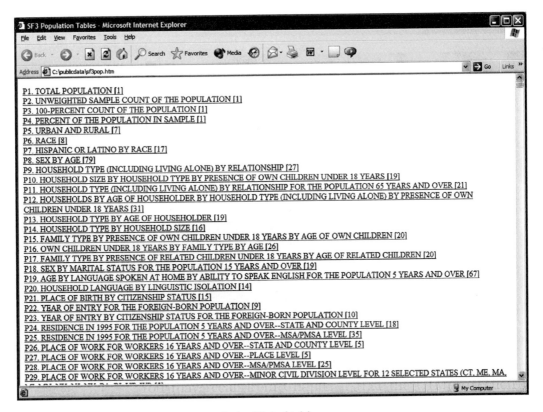

Fig. 7-3. Hyperlinked list of tables.

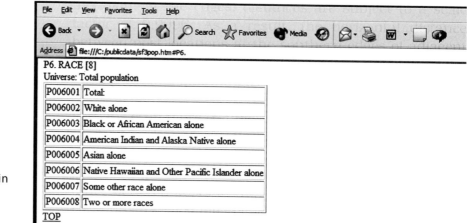

Fig. 7-4.
Variables in
Table P.6,
Race.

# OBTAINING THE DATA

There are two basic approaches to obtaining data for free from the 2000 Census. If you simply need a table for a few areas and a few variables, it is best to go to the Census Bureau's American FactFinder web site (*http://factfinder.census.gov*). That site contains tools for creating queries that extract data from the summary files. It also contains other data sets. However, if you need data for an entire state or the nation, or a large number of enumeration areas, you will probably want to work with the summary file extraction methods described later in this chapter. An example of using the American FactFinder tools is presented in the PowerPoint file for this chapter (see slides 6 through 14). A brief description of that site is given here.

Slides 6–14

The American FactFinder is a web site that contains a series of list boxes that allow you to choose areas and variables. It is particularly useful for a small number of enumeration units and one level below those units. For example, choosing a county and all tracts within it is fairly easy (see figure 7-5). However, choosing all blocks or block groups in a state would be tedious.

You are here: Main ▸ All Data Sets ▸ Data Sets with Quick Tables ▸ Geography ▸ Tables ▸ Results
Census 2000 Summary File 3 (SF 3) - Sample Data, Quick Tables

■ Choose a selection method

| list | name search | address search | map | geo within geo |

Show all geography types | ⓘ Explain Census Geography

■ Select a geographic type
........ ...... Census Tract

■ Select a state
Tennessee

■ Select a county
Blount County

■ Select one or more geographic areas and click 'Add'

All Census Tracts
Census Tract 101
Census Tract 102
Census Tract 103
Census Tract 104
Census Tract 105
Census Tract 106
Census Tract 107          Map It

Add ▼

Current geography selections:
===== Census Tract =====
Census Tract 201, Anderson County, Tennessee
Census Tract 202, Anderson County, Tennessee
Census Tract 203, Anderson County, Tennessee
Census Tract 204, Anderson County, Tennessee          Remove
Census Tract 205, Anderson County, Tennessee
Census Tract 206, Anderson County, Tennessee
Census Tract 207, Anderson County, Tennessee          Next ▶

Fig. 7-5. Choosing tracts for two counties in Tennessee.

Once you have chosen the areas for which you want data, you can then choose the variables to retrieve. Two advantages of the American FactFinder are that you do not have to extract an entire table (although you can) and you can search for variables by subject (see figure 7-6).

While useful, the American FactFinder does have its limitations. You may want variables for many different summary levels and many areas, you may want to study all variables in particular tables, or you may want data for a detailed level (say, block groups) for all counties in a state. The American FactFinder extraction tools are not as useful in these cases (see slide 15). For these tasks it may be easier to obtain the zipped components that make

Slide 15

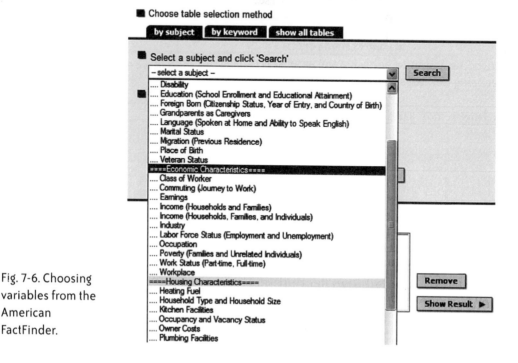

Fig. 7-6. Choosing variables from the American FactFinder.

up the summary files and run them through some form of extraction process. Two approaches are discussed here. The first uses programs that create tables that can be readily linked to shapes from TGR2SHP. The second requires importing data into Access and building queries to retrieve the data you need.

# OBTAINING SUMMARY FILES

As with the data sets discussed in other chapters, the raw data files are available on CD and DVD, or can be downloaded

Slides 16–17

from the Census Bureau's web site. For SF1, the download address is *http://www2.census.gov/census_2000/datasets/ Summary_File_1/* (see slides 16 and 17). For SF3, the URL is the same except that it ends in *Summary_File_3*. You can download each zipped file (40 files for SF1 and 77 files for SF3), or you can download a zipped file that contains all of the zipped files. For example, for Alabama this would be *all_alabama.zip*. Summary files use the following naming convention: *xxyyyyy_ufz.zip* and *xxgeo_ufz. zip*, where:

> *xx* is the two-letter state abbreviation (*us* for the national file),
>
> *yyyyy* is the file number (00001,00002, and so on),
>
> and *z* is the file type (1 for SF1 and 3 for SF3).

Thus, *al00005_uf1.zip* is the fifth file for Alabama for SF1. The *geo* file is the geographic header file. That file contains the information needed to relate the values in the other files to their proper census geography (blocks, tracts, counties, places, ZCTAs, and so on.) The *geo* files have names such as *xxgeo_ufz.zip*. For example, *algeo_uf3.zip* is the geographic header file for Alabama for SF3.

There are several approaches to manipulating the information in the files described here. The Census Bureau has a suggested method for working with Microsoft Access. They also have posted instructions for working with other software systems, such as SAS. Software vendors also distribute programs for extracting tables for the summary files. Your author has written two programs, *SF1toTable* and *SF3toTable*, which are included on the companion CD-ROM. These and the Access approach suggested by the Census Bureau are discussed in the following section.

# SF1toTable and SF3toTable

As with TGR2SHP, I wrote these programs because I thought existing approaches were too time consuming and difficult. Both programs will automatically unzip the files downloaded from the Census Bureau's web pages (listed previously). (If you downloaded the *all_statename. zip* file, you will have to unzip that first.) For each program, there are four things you must do: choose the geographic header file for the summary file you wish to process, choose the tables you wish to extract, choose the summary levels for which you wish to create tables, and set the output directory and file format.

Tables for SF1 and SF3 are grouped into like types. For example, SF1 tables are grouped into the categories Population Tables, Population by Race, Population to the Tract Level, Housing Tables, and Housing by Race Tables. Each group contains a list of available tables that are selected by clicking on the table name and adding it to a list of tables to be extracted. Figure 7-7 presents the population tables for SF1. The third entry in the left-hand window lists *P03. Race [71]*. This is for SF1's table P03. That table contains 71 race-related variables.

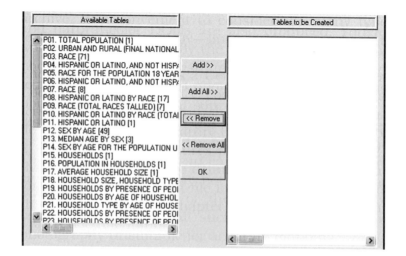

Fig. 7-7. Selecting population tables in *SF1toTable*.

Once the desired tables are selected, the summary levels have to be chosen. The available summary levels are presented as two linked pages of checkboxes. Figures 7-1 and 7-2 are samples of those pages. Finally, the resulting tables can be written in DBF format or ASCII comma-delimited (CSV) format. You must choose one of these formats.

Once you have completed selecting the tables, summary levels, and output format, the program reads the required input files and creates the requested output tables. In addition to the tables you request, a geographic header table is always produced. The resulting tables are organized in directories based on the state and the summary level. Suppose, for example, you wanted to create tables for block groups in Delaware on the C drive of your computer. Those tables will be in a folder named *C:\DE\ Group.*

Slides 18–24

As each table is produced, the programs include information needed to tie the table to its corresponding TIGER polygons. This is best illustrated by way of example. *SF3toTable* was used to extract table P.6 for block groups in the state of Delaware (see slides 18 through 24). In figure 7-4, you can see that that table consists of eight variables containing information on the total population and the totals by certain racial groups. *SF3toTable* created a file named *150P006.dbf.* The *150* signifies the summary level. (You do not have to remember the summary levels, as the output is written in a directory named *Group.*) The *P006* indicates the Census Bureau's name of the table. In addition to the eight variables listed in figure 7-3, the following fields are included in the table:

- ❏ *NAME:* Name of the enumeration unit (e.g., Block Group 2).

- ❏ *FIPSSTCO:* FIPSCODE for the state and county (e.g., 10001).

❑ *TRACT:* FIPSCODE of the Census tract (e.g., 040100).

❑ *GROUP:* Block group number (e.g., 2).

❑ *SFID:* Unique FIPSCODE for this polygon. In this example it would be 100010401002.

The SFID relates directly to the STFID generated by TGR2SHP. Using this table (150P006) and the block group shape generated by TGR2SHP, it is possible—after some manipulation in a GIS program—to map the percentage of persons of two or more races by block group (see figure 7-8).

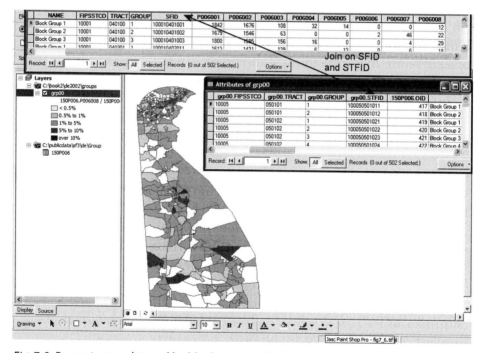

Fig. 7-8. Percentage multiracial by block group in Delaware.

The statement "after some manipulation in a GIS program" is worth considering. In many instances you may want to know the percentage or proportion of the population or households that meet a certain criterion. The Census Bureau usually reports counts. To obtain percentages, you will have to do some manipulation of the data extracted from the summary files.

# Using Access

Before going through the steps required to use Access with the summary files, it is necessary to understand the structure of those files. For any state or for the nation, each file in an SF1or SF3 file set will contain the same number of records, and each record has a unique record number. For example, each of the 77 files in the SF3 for Tennessee data set contains the same number of records. That number reflects the total enumeration units (block groups, tracts, places, and so on) and geographic components used in the state. For both SF1 and SF3, the Geographic Header file contains information that allows you to relate records in the other files to their correct census geography. Without this file, it is impossible to tell if a record refers to the state, a county, a tract, or any other summary level. That is, the Geographic Header file contains information on FIPSCODES.

The Geographic Header file also contains information on Census codes, land and water area, total population, total number of households, and certain legal and statistical area codes. Details of all such fields are in Chapter 2 of *sf1.pdf* and *sf3.pdf*. Records in the numbered files (such as *TN00001.zip*) are related to records in the Geographic Header by their record number. Thus, to extract a table for a particular summary level category (say, Census tracts), you must be able to relate each record in the Geographic Header file to each record in the numbered files.

Slides 25–37

The Census Bureau has built an Access 97 data structure for importing data from the files downloaded for the summary files (see slides 25 through 37). Before importing the data, you must unzip each file. If you are using

Access 2000, you must rename each file to have a *.txt* extension. Once imported you need to build a relationship between each file and the header file (see figure 7-9). To extract particular variables for a specific summary level, you need to build the appropriate query in Access (see figure 7-10). If you want to add a field similar to the SFID field generated by *SF1toTable* and *SF3toTable*, you will need to alter the structure of your query results. This process is also discussed at *http://www.census.gov/support/2000/SF1/Processing.html.*

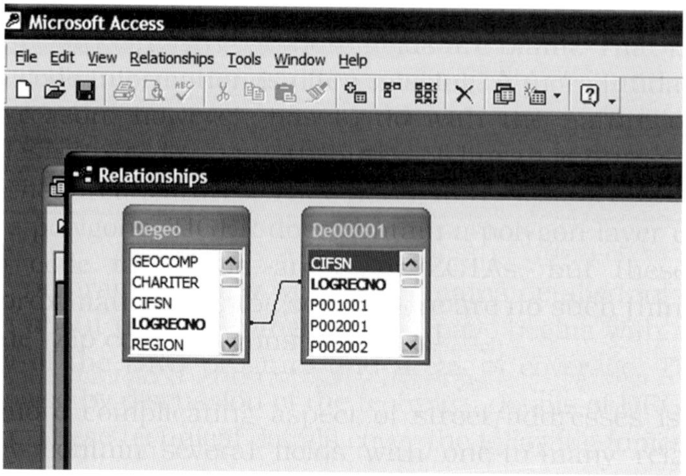

Fig. 7-9. Building a relationship between the Geographic Header File and a summary file.

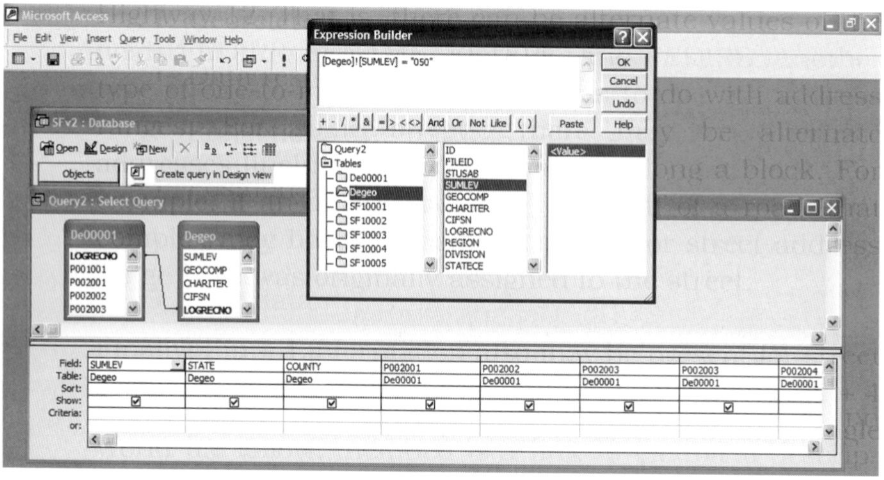

Fig. 7-10. Building a query in Access.

Even if you perform all of these steps correctly, the process outlined previously may not work. The problem is that there are so many records for some states that the database created may exceed the limits of Access. This is true for California using Access 97, which cannot handle files that exceed 1 GB in size. (Access 2000 can take files of up to 2 GB in size.) I must confess to some bias here, but I believe that *SF1toTable* and *SF3toTable* are far easier to use, are much faster, and—in that you have this book—are free. Further, they were designed to be used with TIGER polygons generated by TGR2SHP.

# CHANGES IN RACE AND ETHNICITY DATA

For Census 2000, there was a major change in how race and ethnicity information was collected and reported. There is now much more information about the racial and ethnic heritage of people than was reported in previous censuses. Consider, for example, figure 7-11. It depicts the variable descriptions and variable IDs for table PCT011 from SF1. This table reports at the Census tract level, Hispanic or Latino populations by specific origin. There is quite a bit of detail present in this table. Similar results exist for Africans and Europeans. For American Indian groups, tribal affiliations are reported, whereas for Native Hawaiian and Other Pacific Islanders the islands of origin are reported.

In addition to the increased detail on ethnic origin for certain groups, there has been a major change in how race data are collected. Prior to the 2000 Census it was possible to construct thematic maps of Percent White, Percent Black, and so on. However, prior to the 2000 Census respondents had to choose one (and only one) racial group. This is no longer the case. The data on race now reflects the fact that we are not only a multiracial society but have multiracial and multiethnic heritages. People

**PCT11.  HISPANIC OR LATINO BY SPECIFIC ORIGIN [31]**
Universe: Total population

| | |
|---|---|
| Total: | PCT011001 |
| Not Hispanic or Latino | PCT011002 |
| Hispanic or Latino: | PCT011003 |
| Mexican | PCT011004 |
| Puerto Rican | PCT011005 |
| Cuban | PCT011006 |
| Dominican Republic | PCT011007 |
| Central American: | PCT011008 |
| Costa Rican | PCT011009 |
| Guatemalan | PCT011010 |
| Honduran | PCT011011 |
| Nicaraguan | PCT011012 |
| Panamanian | PCT011013 |
| Salvadoran | PCT011014 |
| Other Central American | PCT011015 |
| South American: | PCT011016 |
| Argentinean | PCT011017 |
| Bolivian | PCT011018 |
| Chilean | PCT011019 |
| Colombian | PCT011020 |
| Ecuadorian | PCT011021 |
| Paraguayan | PCT011022 |
| Peruvian | PCT011023 |
| Uruguayan | PCT011024 |
| Venezuelan | PCT011025 |
| Other South American | PCT011026 |
| Other Hispanic or Latino: | PCT011027 |
| Spaniard | PCT011028 |
| Spanish | PCT011029 |
| Spanish American | PCT011030 |
| All other Hispanic or Latino | PCT011031 |

Fig. 7-11. Summary levels for CD108 Summary File 3.

are no longer forced to choose one race. They can report their mixed-race heritage. If you consider the racial categories in figure 7-4 you will realize that it is possible to create thematic maps of Percent White Alone, Percent Black Alone, and so on. Further, if you consider table P9, it contains categories such as "Asian Alone or in combination with one or more other races."

Thus, it is also possible to create maps, and calculate statistics for Percent with Some White Heritage, Percent with Some Black or African-American Heritage, and so on. In either case, a race alone or a race in possible combinations with other races, the data used to create such maps are not directly comparable to that reported in earlier years. The result is that showing the Percent Change in Non-White Population from 1990 to 2000 is not possible using the data reported by Census 2000. It may be possible to put some upper and lower bounds on what those figures might be, but you would have to assume that the number of people counted as Black or African-American, for example, in 1990 consisted either of those who were Black Alone or those who were Black Alone or in combination with one or more other races.

In either case, assumptions imposed on existing data are impossible to prove and open to debate. Should you come across maps or statistical analyses showing racial change over time, you should in my opinion be quite skeptical. There are two ways of viewing this situation. For some GIS users and students of the Census, this change in reporting presents a real problem. It makes time series analyses of racial change problematic. However, it is much more likely that the data collected in the 2000 Census more accurately reflect America's melting pot heritage. We are a multiracial, multiethnic nation and the new ways of reporting race and ethnicity reflect that diversity.

# OTHER SUMMARY FILE PRODUCTS

The 2000 Census of Population and Housing contains a wealth of information from which the Census Bureau creates many data products. For an introduction to these many products, see *http://www.census.gov/prod/2001pubs/ mso-01icdp.pdf*. Two examples of these additional products are summary files for the 108th Congressional Districts and versions of SF1 and SF3 (called SF2 and SF4) for various racial, ethnic, tribal, and ancestral groups.

Summary files for the 108th Congressional Districts are available at *http://www.census.gov/Press-Release/*

*www/2003/108th.html*. There is a program on the companion CD-ROM, written by this author, for extracting tables from the Summary File 3 version of these files. Called *CD108toTable*, its functions are identical to those of *SF3toTable* (described previously). However, the number of summary levels available in this file is limited to those shown in figure 7-12.

The State level contains only one row: the value of each variable for the entire state. The 108th Congressional Districts level contains geographic components, which are identical to those listed in Table 7-2. Working with the other summary levels, such as CD108 by County, requires some care. As depicted in figure 6-4, congressional districts may cut across any Census polygon that is above the block level of aggregation and below the state level. Thus, the number of entries in a CD108 by County table may exceed the number of counties in a state. For example, Tennessee has 95 counties, but the CD108 by County table contains 107 entries. This is because several counties are split by two or more congressional districts (see slide 38). This can happen at the other levels of aggregation shown in figure 7-12.

Slide 38

SF2 and SF4 contain a subset of the tables found in SF1 and SF3, respectively, but by race, ethnicity, tribal category, or ancestry. The geographic header files of SF2 and SF4 contain a field called the Characteristic Iteration (CHARITER). The value of this field indicates if the data reported in a variable are for the total population (CHARITER = 001) or one of up to 335 subgroups. The subgroups include 132 racial designations, 78 American Indian and Alaska Native tribal categories, 39 Hispanic or Latino groups, and 86 ancestry groups.

A complete list of subgroups can be found in Appendix H of the SF2 and SF4 technical documentation. The number of possible groups, categories, or designations includes combinations of classifications. For example,

**Choose Summary Levels-State Files** ⊠

Choose the areas for which you wish to create tables

Page Options

[Select All on Page] [Clear All on Page]

☐ State

☐ 108th Congressional Districts (CD108)

☐ CD108 by County

☐ CD108 by Tract

☐ CD108 by County Subdivision

☐ CD108 by Place/Remainder

☐ CD108-Consolidated City

☐ CD108-Place in Consolidated City

☐ 108th Congression District by AIANHH

☐ CD108 by AIAN Reservation or Statistical Entity Only (AIANRSE)

☐ CD108 by American Indian (Off Reservation Trust Land Only)Hawiian Homeland (AI(ORTLO)HH)

☐ CD108 by AIANHH Tribal Subdivision Remainder (TSR)

☐ CD108 by AIANRSE TSR

☐ CD108 by AI(ORTLO) TSR

☐ CD108 Alaska Native Regional Corporation

[Cancel] [OK]

Fig. 7-12. Summary levels for CD108 Summary File 3.

Slide 39

the Census Bureau lists 39 American Indian and Alaska Native tribes. For each tribe there are two possible values on any variable: tribe alone or tribe alone or in any combination (see slide 39). An example would be "Cree alone" and "Cree alone or in any combination." The result is 78 (39 x 2) tribal categories.

Reading the definitions of the various categories used in SF2 and SF4 raises interesting questions on the meaning of race, tribe, ethnicity, and ancestry. When, for example, does a designation indicate race and when does it indicate ancestry? There are federal standards on race and ethnicity, as published in the Federal Register Notice on October 30, 1997, in a document titled "Revisions to the Standards for the Classification of Federal Data on Race and Ethnicity," but consider the following sample cases. There is a racial group "White alone" (CHARITER = 002) with Polish (CHARITER = 551) and German (CHARITER = 535) being considered ancestry groups.

Additionally, there is an "Asian alone" (CHARITER = 012) racial group. "Vietnamese alone" (CHARITER = 029) and "Laotian alone" (CHARITER = 024) are also considered racial groups that are subgroups of Asian. Poland and Germany are neighbors, as are Vietnam and Laos. However, one set of neighbors defines ancestral groups, whereas the other defines racial subgroups. According to the guidelines in the Federal Register, race and ethnicity should not be thought of as being based on biologic or genetic characteristics. Instead, race and ethnicity "may be thought of in terms of social and cultural characteristics" (page 58782). The document goes on to caution that the standards for classification "are not intended to be used to establish eligibility for participation for any federal program" (page 58783).

# FUTURE CENSUS RELEASES

When the term *census* comes up, most people think of the Census of Population and Housing. In fact, there are many other products produced by the Census Bureau. The Economic Census is conducted in years ending in 2 and 7, and the 2002 Economic Census will soon be out. (The Census of Agriculture was also conducted in 2002, but by the U. S. Department of Agriculture.)

The American Community Survey represents an effort by the Census Bureau to report detailed, more relevant profiles of American communities more often than just each decade. These and other products mean that there is a steady stream of new census products being produced. They also mean that new versions of TIGER will need to be created to support these products. Knowing how attribute and geographic information can be retrieved, translated, related, and manipulated is important to anyone studying the United States.

# CHAPTER 8

# NATIONAL HYDROGRAPHY DATA

## INTRODUCTION

ALL OF THE PUBLIC DOMAIN VECTOR DATA sets considered in previous chapters had two things in common: they used very simple vector models (points, lines, and polygons) and they were produced primarily by one agency. The National Hydrography Dataset (NHD) is different. The NHD comprises information that defines the surface drainage system for the United States using hydrologic features, including shorelines, rivers and streams, canals, ditches, aqueducts, springs, and the like. NHD is used for modeling point source pollution, potential flood hazards, engineering and construction of bridges, and other hydrologic applications.

The NHD is the product of combining DLG hydrographic files with EPA reach files. Hence, there are at least two metadata files for any NHD data set: those that apply to the DLG information and those that apply to the EPA reach files. Reaches are used to define route systems and regions. Route systems and region data models allow features to be composed of multiple lines or polygons, respectively. Further, they allow lines and polygons to belong to more than one route or region. It is the first property, allowing features to be composed of multiple elements, that is used most often in NHD. Table 8-1 provides an at-a-glance overview of NHD.

**Table 8-1: NHD Overview**

| Component | Description |
|---|---|
| Description | Files that define surface hydrology for the United States. Hydrographic features are modeled using the route systems and region models found in ArcInfo. |
| Format | Usually distributed as zipped ArcInfo workspaces containing route systems and region models. The data is also available in SDTS vector format. |
| Coordinate system | Decimal degrees based on NAD83. |
| Original sources: geographic data | Geographic coordinates are based on DLG hydrographic files. For the conterminous United States and Hawaii, NHD includes complete coverage at the 1:100000 scale. The 1:24000-scale data set is partly done and is in the process of being populated for the United States. For Puerto Rico, the scale is either 1:20000 or 1:30000, whereas the Virgin Islands are at 1:24000. |
| Original sources: reach information | EPA reach files (version 3) are used to assign reach codes and to determine flow direction. |
| Software considerations | Extensions and tutorials for ArcView 3 have been developed for working with the NHD. The ArcGIS 8 model is scheduled for release during the winter of 2004. |
| Other considerations | There are two keys to understanding the NHD: understanding how surface hydrology is conceptualized and understanding how the concepts are implemented using ESRI's route and region models. |
| Software used in this chapter | ArcView 3, NHD ArcView Toolkit. |

The most common form for distribution of NHD data is as zipped ArcInfo workspaces containing composite features based on route systems and regions. This format is referred to as *NHDinArc* and is available for free download via an ArcIMS-managed web site. You can get to it by going to *nhd.usgs.gov* and following the Data link (see slides 1 through 9). In addition to data, the USGS distributes a set of tools for working with NHD data in this format. These include a set of ArcView 3 extensions (discussed in material to follow) for using and maintaining *NHDinArc*. There is also a set of AML routines for appending NHD data sets together. Finally, there are tutorial exercises available from the USGS for working with NHD data sets.

Slides 1–9

There are two major parts to understanding the NHD: you must understand the concepts that underlie the NHD (how hydrologic features and units are conceptualized), and then you need to know how the concepts are implemented. NHD data is more complex than the other data sets described in this book, both conceptually and operationally. The NHD data sets contain many types of features and ancillary attribute tables. This is precisely the type of information that could benefit from an integrated data model in an object relational database, such as the geodatabase.

Slide 10

Indeed, such a database has been developed by Dr. David Maidment of the University of Texas at Austin. For details, go to *www.crwr.utexas.edu/giswr* (see slide 10). As of this writing, the tools for working with NHD data have not been fully ported to ArcGIS. This is scheduled for completion during the winter of 2004. However, there is a Visual Basic Active X DLL available for ArcMap that implements useful functions, such as finding flow directions, calculating network distances, and finding outlets for areas. Developed by Tim Whiteaker, ArcHydro (as it is called) is available at *http://www.crwr.utexas.edu/giswr/hydro/index.html*. This chapter discusses the ArcView 3 extensions for working with NHD, as they are currently available.

Because tools and tutorials exist for *NHDinArc*, the goals of this chapter are to explain the lineage of the NHD data, present its major terms and concepts, illustrate some peculiarities in the data, and indicate the types of functions the NHD data can support. No matter what format and software you use to work with NHD data, you will need to understand these things.

# THE HIERARCHICAL NATURE OF NHD

The NHD uses a hierarchical structure to define hydrologic units and linear features. Just as TIGER divides the United States into states, counties, tracts, block groups, and blocks (each of which forms a nested hierarchy), hydrologic units divide the nation into regions, subregions, accounting units, and catalog units (also called watersheds). Each unit in this hierarchy is assigned a hydrologic unit code, or HUC. In TIGER, successively more detailed units have longer FIPSCODE values, and it is possible to find the state and county a tract is in by looking at the first five digits of its full FIPSCODE. (See the example in Chapter 6 if you need to refresh your memory.)

A similar relationship exists for the levels of HUCs. The hierarchy levels and their names are presented in Table 8-1. The highest-level units, called regions, contain the largest areas and have a two-digit code (see slide 11). The next level of the hierarchy, subregions, has a four-digit code, and so on to the lowest level. At each level, two digits are added to the HUC code. Some accounting units may be coincident with subregions, and unit names may not be unique. The catalog units, or watersheds, are the same units used in the EPA's Surf Your Watershed web site at *http://www.epa.gov/surf* (see slide 12). Figure 8-1 contains watersheds in the Southern Florida subregion.

Slides 11–12

Slides 13–16

**NOTE:** *See see slides 13 through 16 in the PowerPoint file for this chapter for maps of the example areas given in Table 8-2.*

## Table 8-2: HUC Hierarchy

| Hierarchy Level | Level Name | Number of Units | Code Length | Example Code and Name |
|---|---|---|---|---|
| 1 | Region | 21 | 2 digits | 03 – South Atlantic Gulf |
| 2 | Subregion | 222 | 4 digits | 0309 – Southern Florida |
| 3 | Accounting units | 352 | 6 digits | 030902 – Southern Florida |
| 4 | Cataloging unit (watershed) | 2150 | 8 digits | 03090202 – Everglades |

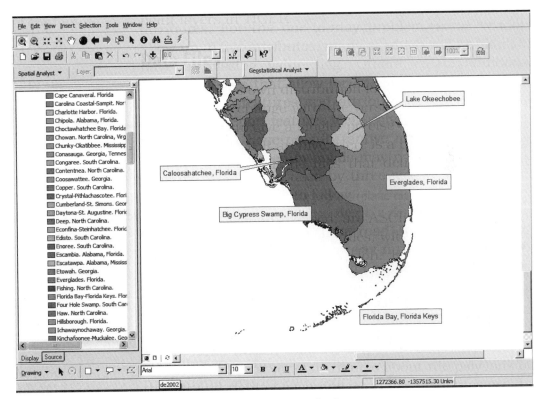

Fig. 8-1. Southern Florida watersheds.

For years, data users had to choose between two main sets of hydrologic vector base data: EPA's River Reach files and USGS digital line graph (DLG) hydrography files. The EPA River Reach files (RF3) contained useful information on river flow direction, network connectivity, and river names, but were lacking spatial completeness and accuracy. The USGS DLG hydrography files contained point, line, and area features, but contained few names and no flow or networking structure or attributes. The NHD was created by the EPA and USGS to provide data users with the best of the RF3 information combined with the best of the DLG information.

The HUC (hydrologic unit code) hierarchy allows data to be organized into nested hydrologic units. You need to consider the features present within those units. The reach files were developed to give each of 3.2 million stream segments, or reaches, a unique code and to establish the flow directions and connections between reaches.

The reach codes, flow direction, and connection information define those features, such as lakes and streams, that comprise the national hydrologic transport network as derived from the 1:100000 hydrography files. When combined with the information in the DLG files, the reach files allow for the delineation of a traceable, measured hydrologic network. Thus, the features in the NHD come from USGS DLGs, whereas the network delineation and reach code ID numbers come from EPA's reach files.

## Features in NHD

NHD contains point, line, and polygon features. The features define the major hydrologic units for the United States. Hydrology is a complex phenomenon, and simple topology does not adequately capture all relationships among these 0D, 1D, and 2D entities. Route and region systems are therefore defined on top of the NHD basic topological features. These are combinations of lines and areas that together define elements of the hydrologic system.

Specifically, the NHD consists of a drainage network, reaches, water bodies, and hydrologic points. As will be seen in material to follow, the relationships between these types of features are maintained through common ID fields. The lines in NHD are digitized so that linear drainage features are digitized in the direction of flow, if it is known, and arcs defining polygons or coastlines are digitized so that water bodies are always on the right. The direction of flow might not be known, for example, in tidal areas. Figure 8-2 shows the arcs defining part of Florida's east coast. Note that the coast lines are digitized so that the water body, the Atlantic Ocean, is to the right of the coastline.

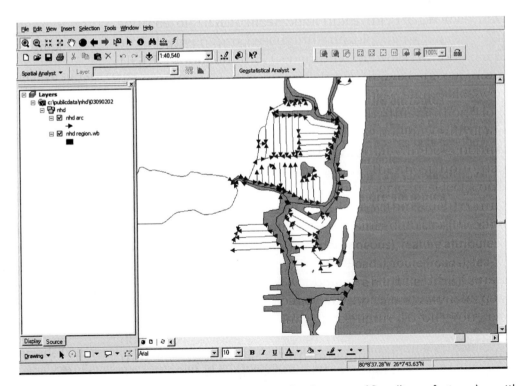

Fig. 8-2. Arcs are digitized in the direction of flow (linear features) or with water bodies on the right (area features).

Features are given a feature code and type. The code, or FCODE, is a numeric value. The type, or FTYPE, is a descriptive label (one for each FCODE value). The relationship between FCODE and FTYPE values is similar to that of *ENTITY_CODE* and MEANING in DLGs. Each FCODE value contains five digits and consists of two parts. The first three digits indicate the feature type, and the last two digits indicate the characteristics of that particular feature. For example, the FCODE for canals and ditches always starts with 336. There are three possible characteristics:

❏ *33600*: Canal/ditch with no attributes

❏ *3360*: Canal/ditch that is an aqueduct

❏ *33602*: Canal/ditch with unspecified attributes

Feature types (and codes) can exist for all topological levels: lines, polygons, and points, and some feature types can exist at multiple levels. Consider, for example, the feature type STREAM/RIVER. At some scales it makes sense to treat elements of this type as lines, but at others it makes sense to treat them as areas. The same is true of dams—their scale will determine if they are line or area features. The feature type RAPIDS can be stored as a point, line, or area feature.

The number of feature types and codes is too large to list here. All possible values can be found in Appendix B of *National Hydrography Dataset Concepts and Contents*, available at *http://nhd.usgs.gov/chapter1/chp1_data_users_guide.pdf*. The various levels (point, line, and area) in which a feature can exist are listed in Appendix A of that document.

Point features are those entities that define hydrologic entities. Examples include springs or seeps, waterfalls, rapids, and gaging stations. These are analogous to entity points (type NE) in SDTS format.

## NHD Reaches

Area and line features can be combined to form reaches. Each reach is assigned a unique reach code, called *RCH_CODE*. The code usually will not change. If it does, say due to extensive changes in the delineation of the hydrographic network, the changes are tracked and the old code is retired. In this sense, the *RCH_CODE* functions very much like the TLID values in TIGER. Each *RCH_CODE* consists of 14 digits. The first eight specify the HUC, otherwise known as the catalog unit or watershed. The remaining six (referred to as the segment number) are for the specific reach in the HUC.

Figure 8-3 illustrates the relationship between reaches and their numbers. In this figure, there are linear and area features with reach codes. Each element in the figure is from the Watts Bar HUC (06010201). The area feature (the Tennessee River) has a *RCH_CODE* of 06010201001647. The labeled streams outside of this area feature have their own segment numbers. Three are labeled in the figure: 000765, 000766, and 000768. The three lines inside the area feature are artificial paths, and they all share the same segment number (and, therefore, *RCH_CODE*) of 001525.

This example illustrates two important points. The first is the existence of artificial paths. This type of feature is used to show the flow path through water bodies, such as lakes or rivers. The second is that multiple features may share the same *RCH_CODE* value. This is true even if the features are not artificial paths.

In addition to a code value, each linear reach is assigned a LEVEL. A feature that drains into a "terminal" water body—such as the oceans, the Gulf of Mexico, and the Great Lakes—is a level 1 feature. A feature that drains into a level 1 feature is a level 2 feature, and so on. Thus, the Tennessee River is a level 3 feature that drains into

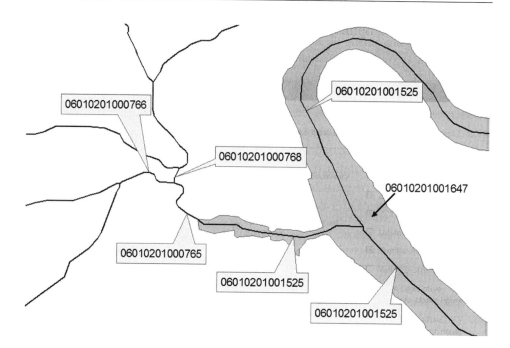

Fig. 8-3. Example reach codes.

the Ohio River (a level 2 feature), which drains into a level 1 feature (the Mississippi River), which drains into a terminal water body, the Gulf of Mexico. Although area reaches do not have levels, the artificial paths that run through them do.

In addition to the *RCH_CODE* and LEVEL, reaches have a reach date attribute, a GNIS identifier, and a name attribute. The last two may be blank. With the name attribute, it is possible to select all reaches that make up a particular feature. For example, figure 8-4 shows all the line features that share the name Knob Creek in HUC 06010201. Note that some of these linear features are artificial paths that track the stream path through a water body.

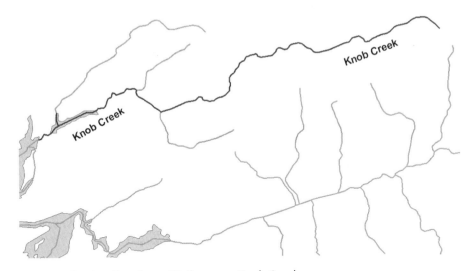

Fig. 8-4. Reaches with the name Knob Creek.

## Flow Relationships

When constructing the reach files, the EPA also developed a flow relationship database. A simplified version of the database structure would be:

```
Reach 1, Reach 2, Direction_Text, Delta Level
```

*Reach 1* and *Reach 2* are values that allow you to identify the reaches in question. The *Direction_Text* can be In, Out, Network Start, Network End, and Non-Flowing Direction. Not all reaches will have an entry in the flow relationship database. For some reaches, their direction may not be known or it may change due to tides. Only those reaches for which the flow direction is known will be in the flow relationship database.

Figure 8-5 illustrates some of these flow relationships. This is near Boyton Beach, Florida, where the Intracoastal Waterway flows into the Atlantic Ocean. The artificial path reach in the Intracoastal Waterway flows into two other reaches. One of those is represented by the artificial path that flows from the Intracoastal Waterway to the Atlantic Ocean. That reach has a network-end flow direction as it connects to the ocean. The shoreline reach has a non-flowing direction.

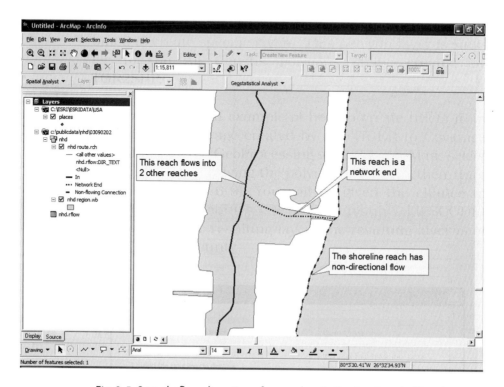

Fig. 8-5. Sample flow directions for reaches in Boyton Beach, Florida.

As noted in the figure, it is possible for a reach to have a flow relationship with more than one other reach. The labeled Intracoastal Waterway reach flows into the continuation of the Intracoastal Waterway and into the Network End reach. The reaches that make up the Intracoastal Waterway are all of level 2, because the Intracoastal eventually connects to level 1 reaches that connect to the Atlantic Ocean. Reaches of the Intracoastal that have a flow relationship have the same level (2), and thus the delta level of that relationship is 0. The relationship between the Intracoastal reach that also connects to the Network End reach has a delta level of 1.

When the Network End reach connects to the shoreline reach, the delta level is assigned a value of –9999.

Slide 17

There are many reaches in the area depicted in figure 8-5 that are not drawn on the map. This is because the figure contains only those reaches with a known flow direction. There are several reaches for which the flow direction changes with the tides. Such reaches will not be in the flow relationship database (see slide 17).

The flow relationship information is very important because it indicates the connectivity and flow direction of the hydrologic network. The structure of the flow relationship establishes which reaches are networked and which are not. A networked reach is one that has at least one entry in the flow relationship database. If a reach is not connected to any other reach, it is said to be unconnected and unnetworked. It does not touch, nor does it share flow with, any other reach.

If a reach connects to another reach but flow direction is unknown, the reach is said to be connected but still unnetworked. That is, it will have no entry in the flow relationship database. If a reach is connected to another reach and the flow direction is known, both reaches are networked. Put another way, if a reach is in the flow relationship database it is connected to another reach and the flow direction between them is known.

# NHDinArc

Previous discussion focused on the conceptual structure of the NHD. In particular, you saw that there are features of various topology (points, lines, and areas), the features have feature types associated with them (e.g., spring/seep, artificial path, lake), and a flow relationship table exists that supplies information on which linear reaches are networked and, if known, on the flow direction between them.

Implementation of these concepts is accomplished using composite features (entities that consist of one or more features) rather than simple topological features. For linear features, the composite features are route systems, whereas area composite features are treated as regions. Composite features allow different route or region layers to share the same underlying coverage. In addition to the usual attribute files and relationships present in route (RAT and SEC files) and region (RXP files) models, the NHD workspaces contain several attribute tables in Info database format. One such attribute table is the flow relationship table discussed previously.

Slides 18–19

An example *NHDinArc* workspace is used to illustrate the various composite features, their attributes, and the related Info database files. The workspace considered is 03090202, the Everglades watershed (see slides 18 and 19). The workspace consists of the following coverages:

❏ NHDDUU

❏ NHDPT

❏ NHD

Slide 20

In addition, there will be a *METADATA* folder in the workspace. The NHDDUU coverage consists of a region theme that contains information on NHD digital update units (DUU). DUUs are polygons that contain features or reaches to which metadata applies. Feature DUUs usually correspond to DLG quadrangles, whereas reach DUUs correspond to sections of an HUC that lie in a specific quadrangle. The region attributes contain names of metadata files in a field named *DUU_NAME*. These files are found in the *METADATA* folder under the *Catalog Unit* folder. Each file has a *.met* extension. The name field indicates the name of the 1:1000000-scale quadrangle. For example, a value *wp2001* indicates that there is a metadata file named *wp2001.met* (see slide 20).

Within that file is a description of the quadrangle for which the DLG hydrography data was extracted. In this case, the *West Palm Beach 1:100000* scale DLG file. In addition, there will be files with names such as *03090202001.met*. You might wonder why there are two different types of metadata files. This reflects the fact that the NHD is derived from USGS DLG hydrography and EPA reach files. The metadata files whose names contain the eight-digit catalog unit identifier contain the EPA-generated metadata. Also included in the region file is the date the digital update unit was created.

The NHDPT coverage consists of landmark points. These are point hydrologic features, such as seeps, springs, gaging stations, and wells. These points are not topological nodes in any line or area features; they are point entities. Topologically, they are similar to the entity points found in SDTS DLGs. In fact, because NHD are derived from DLG hydrography, they are likely to contain many of the same points. The attributes of these points include a feature code, a feature type, a GNIS number, and a name (see figure 8-6). Not all points will have values for GNIS and name, but they will have feature codes with corresponding type descriptions.

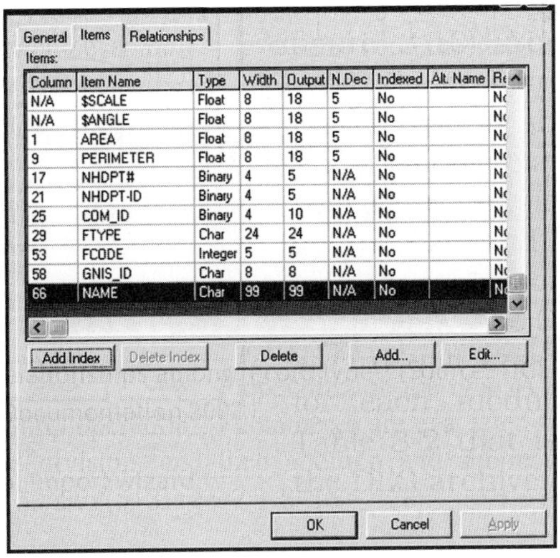

General | Items | Relationships |

Items:

| Column | Item Name | Type | Width | Output | N.Dec | Indexed | Alt. Name | R |
|--------|-----------|------|-------|--------|-------|---------|-----------|---|
| N/A | $SCALE | Float | 8 | 18 | 5 | No | | N |
| N/A | $ANGLE | Float | 8 | 18 | 5 | No | | N |
| 1 | AREA | Float | 8 | 18 | 5 | No | | N |
| 9 | PERIMETER | Float | 8 | 18 | 5 | No | | N |
| 17 | NHDPT# | Binary | 4 | 5 | N/A | No | | N |
| 21 | NHDPT-ID | Binary | 4 | 5 | N/A | No | | N |
| 25 | COM_ID | Binary | 4 | 10 | N/A | No | | N |
| 29 | FTYPE | Char | 24 | 24 | N/A | No | | N |
| 53 | FCODE | Integer | 5 | 5 | N/A | No | | N |
| 58 | GNIS_ID | Char | 8 | 8 | N/A | No | | N |
| 66 | NAME | Char | 99 | 99 | N/A | No | | N |

Add Index | Delete Index | Delete | Add... | Edit...

OK | Cancel | Apply

Fig. 8-6. Attributes of NHDPT coverage.

The NHD coverage consists of line and polygon features. In addition, there may be a node theme for underpasses and overpasses. These could occur, for example, where an aqueduct passes over a stream or canal. Several composite feature types, or subclasses, are defined upon these features. These are listed in Table 8-3. Each subclass is discussed in material that follows.

**Table 8-3: Subclasses Defined on NHD Coverage**

| Underlying Features | Composite Feature Type | Subclass Name | Description | Has COM_ID? |
| --- | --- | --- | --- | --- |
| Lines | Route system | DRAIN | Network element theme | Yes |
| Lines | Route system | RCH | Transport and coast-line reach theme | Yes |
| Polygons | Region system | RCH | Water body reach theme | Yes |
| Lines | Route system | LM | Line landmark theme | Yes |
| Polygons | Region system | LM | Area landmark theme | Yes |
| Polygons | Region system | WB | Water body theme | No |

For several feature types and for the flow relationship table, a common identifier number, or *COM_ID*, is defined. This acts as a database key facilitating relationships between the various themes and the flow relationship table.

# THE NETWORK ELEMENT THEME (DRAIN)

This route system establishes the feature code and type for each linear network element. A linear network element can consist of more than one arc. For example, the artificial path lines in figure 8-4 constitute one element in the drainage theme model. The attributes in the DRAIN route system contain the FTYPE and FCODE for each element, along with the length of the feature in meters. Therefore, you should use this subclass if you want to query or create a thematic map based on feature type.

In addition to the previously discussed attributes, the DRAIN route system contains a field, *RCH_COM_ID*, that allows you to relate each drainage element to the linear reach (transport or coastline) with which it coincides. It also contains a field, *WB_COM_ID*, that when implemented will allow you to associate artificial path drainage elements with the water bodies through which they pass.

# LINEAR REACHES AND FLOW RELATIONSHIPS

The RCH route system contains the EPA-defined reach codes for linear features and their associated fields. Fields include *RCH_CODE* (reach code), *RCH_DATE* (date the code was assigned), *LEVEL*, *GNIS_ID*, *NAME*, *METERS*, and *COM_ID*. Thus, you should use this theme when you wish to select all linear features with a common name, as in *"NAME" = 'Biscayne Bay'*.

**Info Table Properties**

General | Items | Relationships

Items:

| Column | Item Name | Type | Width | Output | N.Dec | Indexed | Alt. Name | Redef |
|--------|-----------|------|-------|--------|-------|---------|-----------|-------|
| N/A | ROWID | OID | 4 | 5 | N/A | No | | No |
| 1 | REL_COM_ID | Binary | 4 | 10 | N/A | No | | No |
| 5 | COM_ID_1 | Binary | 4 | 10 | N/A | Yes | | No |
| 9 | COM_ID_2 | Binary | 4 | 10 | N/A | Yes | | No |
| 13 | SEQUENCE | Integer | 3 | 3 | N/A | No | | No |
| 16 | DIRECTION | Integer | 5 | 5 | N/A | No | | No |
| 21 | DIR_TEXT | Char | 24 | 24 | N/A | No | | No |
| 45 | DELTA_LVL | Integer | 5 | 5 | N/A | No | | No |

Add Index | Delete Index | Delete | Add... | Edit...

Fig. 8-7. RFLOW table structure.

The RCH route system is most often used in conjunction with the flow relationship table, RFLOW. The elements in the RFLOW table are shown in figure 8-7. The RFLOW table contains the *COM_ID*s associated with each pair of reaches that have a flow relationship (*COM_ID_1* and *COM_ID_2*). In addition, each relationship has its own ID value, called *REL_COM_ID*. As described previously, this table is key to understanding the network relationships between reaches.

In addition to the direction and delta level discussed previously, the RFLOW table also contains a field named *SEQUENCE*. In most cases, when two network elements intersect, a new reach code will be defined and the sequence number will be zero. However, if the intersection of a tributary with another reach is deemed insignificant a new reach may not be redefined. (The definition of *insignificant* can be found at *http://nhd.usgs.gov/chapter1/ index.html#_Toc474479805*.) Put another way, a reach is not necessarily created at each confluence. Figure 8-8 illustrates such a case in South Florida. The reach 03090202031870 with *COM_ID* 95006 is the "into" reach for four tributaries. In other words, the four tributaries drain into reach 95006. Their intersections with this reach are given sequence numbers 0, 1, and 2, with the lowest number corresponding to the most upstream confluence.

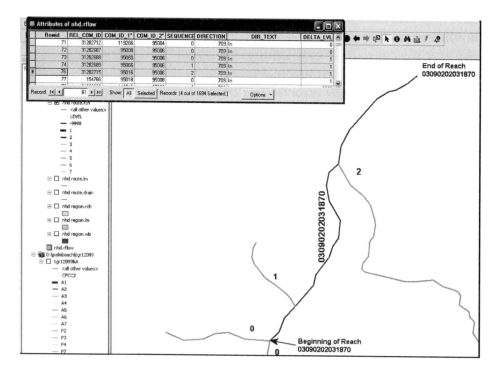

Fig. 8-8. A reach with sequence values greater than zero.

## Water Body Reach Theme

For most of the United States, area-based reaches are assigned to lakes and ponds. Some additional feature types (ice mass, swamp/marsh, and reservoir) have been assigned to features in the state of Washington. Artificial path reaches are often defined to run through water body reaches. See figure 8-4 for an example. However, the linear feature and the area feature it intersects do not share the same reach code. In addition to the *RCH_CODE*, *RCH_DATE*, *COM_ID*, *GNIS_ID*, and *NAME* fields, this theme also contains the area of each feature in square kilometers. The area is stored in the field *SQ_KM*. As with the linear reaches, you can select features based on name, as in the query *"NAME"* = *'CLEAR LAKE'*.

## Landmark Route Theme

Linear landmarks are stored in this theme. Fields for this theme include a *COM_ID*, *METERS* (the length of the feature), *FTYPE*, *FCODE*, *GNIS_ID*, and *NAME*. Examples of these types of features are *GATE*, *WALL*, *DAM/WEIR*, *LOCK CHAMBER*, and *NONEARTHEN SHORE*. For example, in the Watts Bar HUC the Tennessee Valley Authority's Watts Bar Dam is a found in this theme.

## Area Landmark Theme

These are landmark features that are large enough to be modeled as areas. An example would be Fort Loudon Dam in the Watts Bar HUC. Although this is also a TVA dam, it is large enough to belong to the region model for landmarks (see slide 21). Other types of features found in this theme include *SUBMERGED STREAM*, *LOCK CHAMBER*, and *DAM/WEIR*. Attributes for this theme include *COM_ID*, *FTYPE*, *FCODE*, *SQ_KM*, *GNIS_ID*, and *NAME*. In addition, there are fields for elevation (*ELEV*) and the stage at which the elevation applies (*STAGE*).

Slide 21

## Water Body Theme

It is possible for water bodies to be neither a reach nor a landmark. In the Everglades watershed, the Everglades themselves would be such an example. This theme includes a field, *RCH_COM_ID*, that allows you to associate water bodies with corresponding area reach theme features. The common ID field is, curiously enough, not named *COM_ID*. It is named *WB_COM_ID*. As mentioned in the discussion of the network element theme (DRAIN), future releases of NHD should make it possible to link the water body theme with the artificial paths. This will make it possible to find all artificial paths that flow through a swamp, for example. Other fields in the WB region theme include *FTYPE*, *FCODE*, *ELEV*, *STAGE*, *SQ_KM*, *GNIS_ID*, and *NAME*.

# INFO TABLES

In addition to the standard attribute tables found with coverages, route systems, and regions, there are several other attribute tables found in each workspace. You have already read about the flow relational table (RFLOW). Other files are listed in Table 8-4.

**Table 8-4: Other Tables in NHD Workspaces**

| Table Name | Scripting Name | Description |
|---|---|---|
| Reach Flow | Rflow | Used to establish connectivity, flow direction, and sequence order of reaches. |
| Feature Code | Fcode | Contains five-digit feature codes, the types they represent (e.g., stream/river, dam/weir, canal), along with 26 other sub-attributes (e.g., construction materials, flow status). |

| Table Name | Scripting Name | Description |
|---|---|---|
| Polygon-Region Cross Reference | Rxp | Used to relate polygons to NHD features. A single polygon may belong to more than one feature. |
| Reach Cross Reference | Rcl | Used to track changes in reach codes as NHD is edited. Contains codes for reaches that have been deleted, added, or redelineated. |
| Feature Relationship | Frel | Relates arcs to drainage features and polygons to water body features. |
| Status | status | Tracks changes to features and relationships, such as adds, deletes, modified attributes, modified spatial elements, and validated entities. |
| Feature/Reach Digital Update Unit Association | duu2fea | Relates each feature or reach to its update unit. |
| Relate | rel | An ArcInfo SAVE RELATE generated table of relationships between NHD tables. |
| Metadata tables | *.met | Metadata for each digital update unit—usually 1:100000 DLG quads and reach file cataloging unit. |
| NHD ArcView 3 tools *Openme* file | openme.txt | File used by ArcView 3 NHD tools to load in an NHD workspace. |

The feature code table (FCODE), the Polygon-Region Cross Reference table (RXP), the Reach Cross Reference table (RCL), the Feature Relationship table (FREL), the Status table (STATUS), the Digital Update Unit Association table (DUU2FEA), and the Relate table (REL) are all Info tables. The metadata tables and the *Openme* file are text files. The metadata tables were discussed previously.

The *Openme* file is used by the ArcView extension developed for the NHD.

# TOOLS FOR WORKING WITH NHD

As mentioned at the start of the chapter, the USGS has created extensions for working with NHD data in ArcView 3 (see slide 22). The extensions contain custom menus and tools for many tasks, including extracting and loading NHD workspaces, tracing upstream and downstream of a point, and adding event themes. For example, figure 8-9 shows those reaches that are within 5 kilometers upstream of where Knob Creek enters the Tennessee River. You can also convert NHD coverages to shapes. In addition, the relationships between the various tables and spatial elements are created automatically. Screen shots of the extensions are in the PowerPoint slides for this chapter (see slides 23 through 37).

Slides 22–37

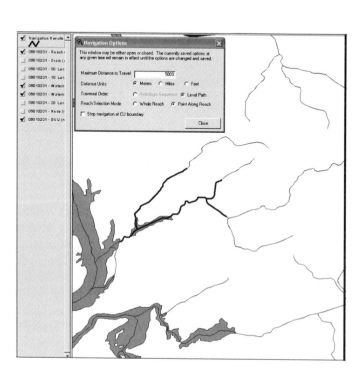

Fig. 8-9. Linear reaches that are within 5 km of where Knob Creek enters the Tennessee River.

*NOTE: Tutorials for working with the extension are available at* http://nhd.usgs.gov/tutorials.html. *In addition to the tutorials, a Workstation ArcInfo AML for appending NHD workspaces is available at* http://nhd.usgs.gov/tools.html#append.

As should be evident from the text, NHD workspaces are complex. Appending two or more workspaces requires updating routes and regions, removing duplicate features, and updating the related tables. If you wish to edit an NHD workspace, you must be sure to use the proper editing tools. Simply editing arcs will not properly maintain the route and region structures. Whatever GIS software you use, you should be familiar with linear referencing before editing these workspaces.

Slides 38–41

As with any large data set, the NHD is not without errors and inconsistencies (see slides 38 through 41). Perhaps the most troubling of problems occurs at the boundaries of digital update units. There is inconsistent conversion of polygons to NHD region features. Figure 8-10 illustrates such a case. Water polygons corresponding to a lake are present on one side of a DUU boundary but absent on the other. This occurs even though the polygons exist in the DLGs that correspond to each DUU.

Fig. 8-10. Inconsistent conversion of DLG polygons to NHD water bodies between DUUs.

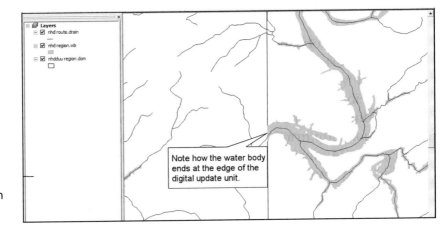

# FUTURE NHD RELEASES

The way in which NHD data sets are maintained, stored, and manipulated may change in the coming years. The relationships between the various tables, the use of composite features, and the need to work with various units of differing topologic levels such as reach linear and area features naturally lend themselves to object relational data models. It is likely that NHD data will be ported to the geodatabase model or similar object relational structure. For the geodatabase model to become widely accepted, the tools developed in the ArcView 3 NHD extensions and the Append AML will have to be rewritten to work with ArcObjects.

# WEB SITES AND DOCUMENTATION

There are several web sites associated with the NHD. These are accessible from the NHD Technical Reference web page (see figure 8-11). The NHD Fact Sheet is available at *http://erg.usgs.gov/isb/pubs/factsheets/fs110699.html* (see slide 42). The technical standards can be found at *http://rmmcweb.cr.usgs.gov/public/nmpstds/nhdstds.html.* As with other USGS technical documentation, this page contains links to PDF documents (see slide 43). There are several links to documents that pertain to the NHDinArc implementation.

Slides 42–43

The primary reference for the NHD is the Concepts and Contents document. The HTML version of the document (it is also available in PDF format) contains a hyperlinked table of contents. There are several illustrations in the Concepts and Contents document that explain the various rules used in the NHD, and deviations from those rules (see slides 44 and 45). There are sections on the rules used to create the NHD, and exceptions to those rules. There is even a section titled "Appendix G

Slides 44–45

Peculiarities" that discusses aspects of the NHD that "tend to be irritating, but they do not undermine the usefulness of the NHD." Appendix G also contains steps that can be taken to address such irritations on a case-by-case basis.

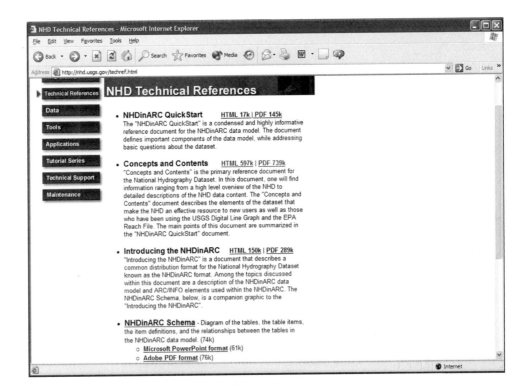

Fig. 8-11. The NHD Technical References page.

# INDEX